FORMAL STRUCTURE OF ELECTROMAGNETICS

FORMAL STRUCTURE OF ELECTROMAGNETICS

General Covariance and Electromagnetics

E. J. POST
Physics Department
University of Houston (ret.)

DOVER PUBLICATIONS, INC.
Mineola, New York

Published in Canada by General Publishing Company, Ltd., 30 Lesmill Road, Don Mills, Toronto, Ontario.

Published in the United Kingdom by Constable and Company, Ltd., 3 The Lanchesters, 162–164 Fulham Palace Road, London W6 9ER.

Bibliographical Note

This Dover edition, first published in 1997, is an unabridged and unaltered republication of the work first published by North-Holland Publishing Company, Amsterdam, The Netherlands, in 1962.

Library of Congress Cataloging-in-Publication Data

Post, E. J.
Formal structure of electromagnetics : general covariance and electromagnetics / E. J. Post. — Dover ed.
p. cm.
Originally published: Amsterdam : North-Holland Pub. Co. ; New York : Interscience Publishers, 1962. (Series in physics)
"Unabridged and unaltered republication"—T.p. verso.
Includes bibliographical references and index.
ISBN 0-486-65427-3 (pbk.)
1. Electromagnetic theory. 2. Transformations (Mathematics)
I. Title. II. Series: Series in physics.
QC670.P77 1997 97–16159
530.14'1 — dc21 CIP

Manufactured in the United States of America
Dover Publications, Inc., 31 East 2nd Street, Mineola, N.Y. 11501

This book is dedicated to the memory of my father who lived a life firmly anchored in the values of reality and integrity. May some of his philosophy be present in the pages of this text.

PREFACE

This book presents an explicit treatment of the principle of general covariance as applied to electromagnetics, de-emphasizing the customary close association between general covariance and relativity.

A certain completeness in the coverage of coordinate transformations is claimed. Special attention has been given to the transformation theory of physical fields so as to provide the *a priori* knowledge essential to the use of Lagrangian methods. A mathematical parallel with the theory of contact transformation in Hamilton-Jacobi theory has not been deemed justified but a few preliminary steps have been taken to stimulate investigations in that field. The ideas evolved in the theory of potentials may help to shed light on the mutual relationships between gauge and contact transformations.

I became closely acquainted with the subject matter of covariance through Prof. J. A. Schouten. His work, as well as Prof. D. van Dantzig's work on electromagnetics, has greatly influenced this book. The text can in a sense be regarded as a physicist's interpretation and compilation of their more mathematically oriented investigations. In this regard I was helped by many enlightening discussions with Dr. G. Weinreich (now at Ann Arbor, Mich.).

Many other colleagues have given me the benefit of their comments on various parts of the manuscript. I am particularly grateful to Prof. S. A. Wouthuysen, whose constructive criticism helped me clarify cryptic passages, correct several minor errors, and overcome a few major deficiencies.

The first half of the manuscript was roughed out while I was with the Bell Telephone Laboratories in New Jersey. The entire manuscript was completed in December 1960.

It is a pleasure to acknowledge the encouraging discussions with Dr. A. Walther during the early months of 1961 when both of us were with Diffraction Ltd., Inc., in Bedford, Mass. The staff of the Air Force Cambridge Research Laboratories kindly assisted in the editorial details of proofreading. Dr. E. F. Bolinder gave the proofs a thorough second reading while checking the formulae. Messrs P. T. Bailey (MIT) and J. Beall (U.S. Wheather Bureau) kindly helped me through the first reading.

February 1962 E. J. Post
Bedford, Massachusetts

In 1996 we got word that S. A. Wouthuysen, emeritus professor of physics in the University of Amsterdam, had passed away. This book came about as a result of a major initiative on his part to help preserve some of the physics-related endeavors of the Schouten school. This is a belated and overdue expression of recognition to both as mentors and friends, for their patience and above all for their openness of discussion.

April 1997 E. J. P.
Westchester, California

TABLE OF CONTENTS

THE FIELD EQUATIONS

THE MEDIUM

LIST OF SYMBOLS

(Latin and Gothic symbols are listed together,
followed by Greek and other symbols)

	Three dimensions	*Four dimensions*
A	action	action
$\boldsymbol{A}$	vector potential	
A_λ		four-potential
$A_\lambda^{\lambda'}$	Jacobian transformation matrix	Jacobian transformation matrix
$A_{\lambda'}^{\lambda}$	inverse Jacobian transformation matrix	inverse Jacobian transformation matrix
$\boldsymbol{B}$	magnetic induction	
$B_{\lambda\nu}$	magnetic induction on general coordinates	
c	light velocity	
$\mathfrak{c}^\lambda$		four-vector of charge and current density, general coordinates
$\mathrm{d}\mathfrak{f}$		volume element of space-time
$\mathrm{d}\boldsymbol{r}$	vector line element	
$\mathrm{d}s$		scalar "line" element of four-space
$\mathrm{d}x^\lambda$	vector line element, general coordinates	vector line element, general coordinates
$\mathrm{d}\sigma$	surface element	
$\mathrm{d}\tau$	space volume element	
∂_λ	partial differentiation with respect to general space coordinates	partial differentiation with respect to general space-time coordinates

	Three dimensions	*Four dimensions*
$\boldsymbol{D}$	electric displacement	
$\mathfrak{D}^{\lambda}$	electric displacement on general coordinates	
$\boldsymbol{E}$	electric field	
$\mathfrak{E}_{\lambda}$	electric field on general coordinates	
$\mathscr{E}$	energy density	
$\mathfrak{f}$		domain of space-time
$\boldsymbol{F}$	Fitzgerald vector	
$F_{\lambda\nu}$		six-vector of electric field and magnetic induction (general coordinates)
$\mathfrak{F}_{\lambda}$		four-vector of rate of change of energy density, and force density (general coordinates)
$\mathfrak{G}^{\lambda\nu}$		six-vector density of electric displacement and magnetic field (general coordinates)
$\hbar$		Planck's quantum of action (divided by 2π)
$\boldsymbol{H}$	magnetic field	
$\mathfrak{H}^{\lambda\nu}$	magnetic field on general coordinates	
l	length	
L	Lagrangian of a system	
$\mathscr{L}$		Lagrangian density
$\underset{u}{\pounds}$	See next column	Lie derivative with respect to displacement field u^{λ}
$\underset{0}{\mathscr{L}}$		Lagrangian density for a nonconducting medium
m	mass	
$\boldsymbol{M}$	vector of magnetic polarization density	

	Three dimensions	*Four dimensions*
$\mathfrak{M}^{\lambda\nu}$		six-vector of electric and magnetic polarization density (general coordinates)
$\mathfrak{N}^{\lambda}$		four-vector density of particle flow (general coordinates)
p_{λ}		general momentum vector (general coordinates)
$\boldsymbol{P}$	vector of electric polarization density	
q		charge
$R_{\nu\sigma\kappa}{}^{\lambda}$		Riemann-Christoffel tensor
$\boldsymbol{s}$	vector of current density	
$\mathfrak{s}^{\lambda}$	vector of current density (general coordinates)	
$\mathfrak{T}_{\nu}{}^{\lambda}$		tensor of energy and momentum density (general coordinates)
t	time	
u	phase velocity	scalar line parameter
u^{λ}		time-space displacement vector
$\boldsymbol{v}$	velocity vector	
x^{λ}	spatial coordinates $\lambda = 1, 2, 3$	space-time coordinates $x^0 = t$
$\overset{0}{Y}$		free space admittance
$Z^{\lambda\nu}$		Hertz-Fitzgerald six-vector
$\tilde{\gamma}_{lk}$	pseudo tensor of the directive effect	
$\Gamma^{\kappa}_{\lambda\nu}$		coefficients of linear displacement or Christoffel symbols
Δ	determinant of Jacobian transformation matrix	determinant of Jacobian transformation matrix

	Three dimensions	*Four dimensions*
ε	dielectric permittivity, $\varepsilon = \underset{0}{\varepsilon}\underset{r}{\varepsilon}$	
$\underset{r}{\varepsilon}$	relative dielectric permittivity	
$\underset{0}{\varepsilon}$	free space dielectric permittivity	
ε_{lk}	permittivity tensor (Cartesian coordinates)	
$\varepsilon^{\lambda\nu}$	dielectric permittivity (general coordinates)	
μ	magnetic permeability, $\mu = \underset{0}{\mu}\underset{r}{\mu}$	
$\underset{r}{\mu}$	relative permeability	
$\underset{0}{\mu}$	free space or absolute permeability	
$\mu_{\lambda\nu\sigma\kappa}$	permeability on general coordinates	
π	Hertz vector	
ρ	charge density	
φ	scalar potential	
χ_{lk}	inverse permeability (Cartesian coordinates)	
$\chi^{\lambda\nu\sigma\kappa}$	inverse permeability tensor on general coordinates	constitutive tensor (general coordinates)
$\underset{0}{\chi}^{\lambda\nu\sigma\kappa}$		free space constitutive tensor (general coordinates)
∇_λ		covariant derivative with respect to space and time coordinates

CHAPTER I

INTRODUCTION

1. MOTIVATION AND OBJECTIVES

The objectives of this monograph are twofold. It is intended to present briefly the formal aspects of electromagnetism but in addition it will be attempted to present a heuristic justification of the presently accepted forms.

Physics, as we know, is not as ideal a playground for formalisms as mathematics for the very reason that the criterion of logical consistency alone is not sufficient for a physical theory. Apart from being logically consistent a physical formalism should offer a unique and unambiguous picture of a field of physical experience, which means that it should give a reasonably objective and quantitatively fitting description of a class of physical phenomena.

A good physical theory, which is not just an *ad hoc* mathematical description of some isolated phenomenon, should have the property that it is sort of self propelling, able to coast along on its own power without too many changes to its fundamental machinery.

Physics is badly in need of principles which help to single out the physically most significant structures among the multitude of mathematical possibilities which are admissible for the description of physical phenomena. In other words, to bridge the gap between physics and its mathematical image one may attempt to tighten the rules for expressing physics in terms of mathematics. We shall endeavor to do so, guided by principles of a knowledge-theoretical nature.

It is obviously not the pretence of the present monograph to do so for all field theories of physics. As regards the present state of the art, however, it is believed to be of interest to review and complete some of the general principles underlying physics' oldest field theory i.e., the theory of the electromagnetic field. There is hardly any other field theory in classical or modern physics that has had an impact comparable to the theory of electromagnetism and which at the same time has a reputation of an almost unquestionable general validity.

It seems to be safe to assert that macrophysically, electromagnetics is a correct and reliable theory while the theory is still useful and to the point in many fields of microphysics.

The previous statements should not be interpreted in a sense that the theory can give all answers in physics and technology within the confines of macroscopic applications. The statement rather substantiates a strong confidence in the general validity of the Maxwell field equations. Any special problem, however, requires a specification of the electromagnetic properties of the medium under consideration as expressed in the constitutive equations.

It is the general feeling that discrepancies between theory and experiment have been due to inadequacies of the constitutive equations rather than to inadequacies of the Maxwell field equations themselves.

In the following chapters we will assume that the reader has a reasonable working knowledge of Maxwell theory and that there is some familiarity with Lorentz transformations and Minkowski's four-dimensional representations of electromagnetics.

It is often remarked that Minkowski's four-dimensional form of Maxwell's theory did not add any essentially new physical content to the theory. In all fairness one might have hardly expected it to do so, because for all practical purposes, electromagnetism was already a firmly established and complete theory by the time of Minkowski's contribution.

The pro's and con's of Minkowski's contribution cannot be decided on the basis of its immediate usefulness in the form of striking new effects. The latter viewpoint reveals a deplorable lack of appreciation for a scholarly effort.

The real value of Minkowski's contribution is not that it revealed a striking effect but that it uncovered an important new aspect of an existing and well-established theory.

The objectives of the next following chapters will be to study the formal aspects of Minkowski's four-dimensional representations of electromagnetics, imposing the condition of covariance for general real and continuous space-time transformation.

Any such study can be completed in an unambiguous way only if precise statements are being made about the mathematical transformation behavior of the relevant physical fields with respect to such general space-time transformations.

The transformation behavior of physical fields will be discussed in chapter 2 on the basis of a dimensional argument which suggests the use of action and charge as the two invariant physical units, whereas length and time are solely associated with the transformation behavior of the fields.

The transformation behavior of the electromagnetic fields ensuing from the considerations in chapter 2 entails the natural invariance of the Maxwell field equations discussed in chapter 3. Natural invariance here implies that the Maxwell equations can be written in an invariant form independent of the structural properties of the space-time continuum, such as the so-called metric of space-time.

The natural invariance of Maxwell's equations, incidentally, helps to bridge the gap between the material medium and the medium of matter-free space. It is usually taken for granted that the constants characterizing a material medium (permittivity and permeability) occur in the constitutive equations and not in the Maxwell equations. For reasons of convenience it has been tacitly accepted for a long time that this is not true for the medium of matter-free space.

The natural invariance of the field equations enables one to treat all electromagnetic media on the same footing, i.e., the properties of the medium do not occur in the field equations but in the constitutive equations only. The properties of the medium known as matter-free space will necessarily be associated with the structure (metric) of the space-time continuum. Hence covariant derivatives which depend on the metric do not belong in the covariant form of the Maxwell equations if we like to adhere to the principle of a functional separation of constitutive equations and field equations. Natural invariance reduces the Maxwell equations to purely topological restrictions on the behavior of the electromagnetic field regardless of the medium.

In the chapters 4 and 5 the variational and Lagrangian aspects of electromagnetics will be discussed. The uniqueness of the Lagrangian procedure hinges on the transformation behavior of the electromagnetic potentials established in chapter 2. It leads quite naturally to the field equations and the stress-energy relations for an unspecified medium. The latter point is helpful to settle a few undecided issues of the theory, i.e., the form and the interpretation of the stress and energy relations in an arbitrary medium.

The first five chapters in principle complete the discussion of the formal structure of electromagnetics. The last four chapters discuss specific examples of so-called admissible Lagrangians pertaining to anisotropic nonuniform and nonreciprocal media.

The real strength of a tight formal approach, however, can be most adequately demonstrated in the case of matter-free space with a gravitational field. In this case there seems no realistic possibility to construct a microphysical model to guess the correct form of the constitutive equations. There is no alternative but the use of a tight formal approach as a last recourse.

The propagation of electromagnetic waves in matter-free space in the presence of a gravitational field is discussed in chapter 9. The geometric optic approximations of the solutions of the wave equations provide an interesting opportunity to study the relations with the relativistic postulate about light rays propagating along null geodesics of the space-time continuum.

2. THE PRINCIPLE OF COVARIANCE

It is an important characteristic of scientific theory that it aims at eliminating irrelevant subjective aspects from a field of human experience. In doing so it creates a possibility to focus on the important observations.

The covariance principle in physics is a very typical example of this objectivation procedure which aims at an elimination of the subjective feature of the space-time frame of reference in the formulation of physical relations. In other words, the covariance principle (or the principle of form invariance of physical laws) emphasizes a formulation of physical laws so that observers in different positions and in different states of motion, including accelerated motion, can use these same laws and have unambiguous means of correlating their observations.

It is usually tacitly assumed that the covariance principle made its entry in physics with the era of relativity. A little further thought, however, shows that covariance principles in a somewhat more restricted form have always played a role in physics. The point is that relativity, for the first time, recognized the essential knowledge-theoretical aspects in the principle of covariance. Before the era of relativity, covariance did play a role but on a much more intuitive level.

A somewhat oversimplified example may help to clarify the nature of the covariance principle. Suppose an observer makes time and position measurements of a material object sliding down the ideally smooth surface of a tilted table. He will then arrive at a set of well-known conclusions about the mutual relations of acceleration, velocity, etc. In addition, the observer will notice that the magnitude of the measured values will depend on the angle of tilt of the table in the form of a cosine-law.

The occurrence of this cosine-law now can be regarded, for the time being, as a typical feature of the experimental arrangement with the tilted table. This represents a subjective but correct recording of the phenomenon.

Now suppose the observer turns theorist, having been made suspicious by other but similar observations. He tries to dissociate the cosine-law from the specific tilted table experiment. Say he identifies the cosine-law observation with the vectorial properties of force, acceleration etc. In the latter case, the observer clearly invokes the issue of a restricted covariance in his effort to come to a more objective and general evaluation of his findings.

The above example shows two important requirements for a meaningful application of the covariance principle. First of all it is necessary that one clearly states the nature of the set of transformations which need be considered. Secondly it is necessary to establish the physically meaningful transformation behavior of the physical quantities involved for any transformation element of the set.

The set of transformations under consideration for the example of the tilted table is simply the group of rotations in space of mutually stationary frames of reference. It is this group of transformations that has led to the system of vector analysis and Cartesian tensors or dyadics for the formulation of the basic laws of physics. Gibbs' vector analysis is one of the important standardized means of mathematical communication in physics which clearly invokes a restricted covariance in space.

In the middle of the 19th century, the question arose whether the linear Cartesian frames could be extended throughout space. It was felt that this might be a somewhat audacious assumption and that an experimental check might be desirable to probe the extent of local applicability of Cartesian frames. Actually, the latter point was carefully checked by Carl Friedrich Gauss on occasion of a general topographical survey of Germany. Gauss optically measured the angles of a large triangle from three mutually visible high spots in Germany and found that the angles added up to 180° within the limits of experimental error. This familar criterion for the conditions of Euclidian geometry seemed to decide the issue that for all practical purposes the group of rotations should be adequate to describe the spatial aspects of physics.

Although it had been decided that no important quantitative physical issue was involved in the assumption of having Cartesian coordinates throughout space, investigations about the form of physical laws on general spatial coordinates continued for purely mathematical reasons. Particle mechanics was one of the first physical disciplines that was subjected to this treatment by Lagrange, Hamilton and Jacobi. Riemann proceeded to develop the geometry of non-Euclidian space.

Along with the (time independent) spatial covariance aspects, definite ideas developed about the behavior of physical laws with respect to mutually translating spatial frames of reference. Hence thus far physics had developed along two disconnected covariance principles: a principle of spatial covariance and a principle of "motional" or "time" covariance.

The most conspicuous aspect of what has just been called tentatively "the principle of time covariance," is the invariance of mechanical laws under uniform translations (Galilei transformations).

A new break-through was effected at the turn of the century when Lorentz remarked that Galilei transformations did not play the same role in empty space electromagnetics as in mechanics. He pointed out that a transformation could be found which was associated with electromagnetics in a similar manner as Galilei transformations are associated with mechanics. This transformation, now known as the Lorentz transformation, accounts for the negative results of the Michelson-Morley experiment and for the first time treated time on a similar footing as spatial coordinates.

The paradoxical situation of having different invariance aspects associated with mechanics and electromagnetics was resolved by Einstein's successful effort to modify mechanics to conform with the Lorentz picture. The merger of space and time covariance within the framework of linear transformations was further completed by Minkowski.

General Lorentz transformations are known to be composed of uniform translations (special Lorentz transformations) and (time independent) rotations of the spatial frame of reference†. Though it had been decided already, on basis of Gauss' experiment, that the group of rotations seems adequate for the spatial description of physics, a restriction to uniform translations, as remarked by Einstein, would be undesirable because it would exclude the treatment of accelerated systems. But then, because of covariance, admitting nonuniformity in the time, it would necessitate considering nonuniformities in the spatial domain no matter how small this inhomogeneity might be for all practical purposes as suggested by Gauss' experiment.

These considerations led Einstein to institute a program to investigate the form invariance of physical laws with respect to general space-time transformations. This program has resulted in the theory of general relativity with a theory of gravitation as its most conspicuous part.

† Excluding coordinate and time inversions. See for instance J. L. SYNGE [1956] chapter 4 for properties and restrictions on Lorentz transformations.

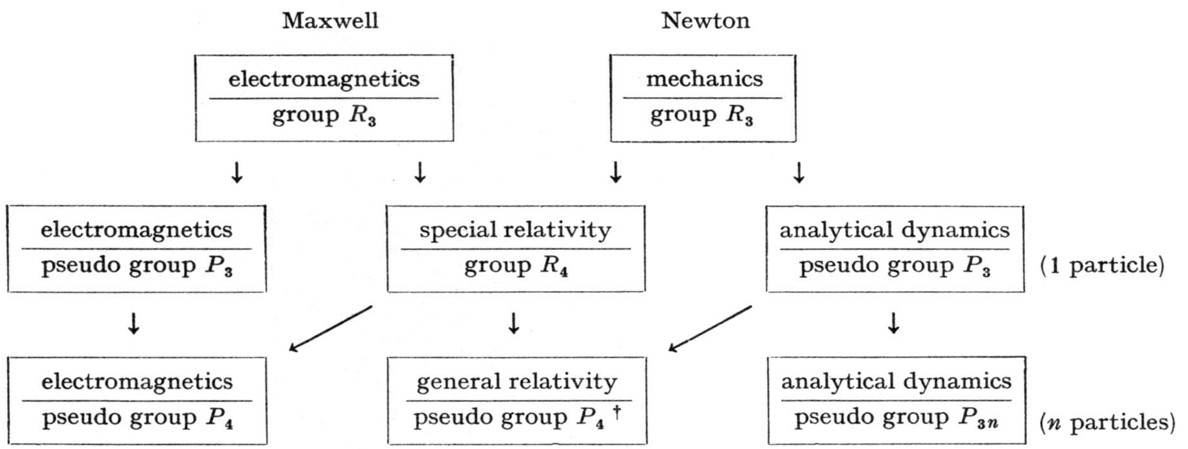

Fig. 1. The evolution of the covariance principle; the symbols R denote "orthogonal" transformations, the symbols P denote general coordinate transformations. The subscripts denote the number of relevant coordinates.

† Einstein really restricted himself to the unimodular subset of P_4 (Jacobian unity). (See chapter 3, section 1.)

We have tried to illustrate the evolution of the covariance principle by means of the block diagram in Fig. 1. Every block in Fig. 1 represents a topic of physical theory and has associated with it the transformation group that characterizes its covariance aspects. The symbol R_3 denotes the group of rotations in space and R_4 denotes the Lorentz group, while P_3 denotes the "pseudo" group of general coordinate transformations in space, and P_4 the "pseudo" group of general transformations in space-time. The term "pseudo" group is customary for the set of general coordinate transformations, because sometimes two successive transformations do not lead to a new transformation element that still belongs to the set. A. Nyenhuis, "Theory of the Geometric Object" (Thesis, Amsterdam 1952), considers the set of general coordinate transformations as a "groupoid" according to Brandt (Mathematische Annalen 96 (1926) 360). Nyenhuis shows that the groupoid of general coordinate transformations is the direct product of a simple so-called minimal groupoid and a group. We may hold on here to the somewhat old-fashioned terminology pseudo group instead of groupoid, because the groupoid may be considered as a group for all practical purposes to be considered in this text.

With due allowance for differences it might be said that Fig. 1 illustrates a classification of physics similar to Felix Klein's classification of geometry.

The topics denoted in the two blocks in the lower left corner of Fig. 1 are primarily the subject matter of this text. Electromagnetics covariant for the pseudo group P_3 is discussed in chapter 7; the other chapters are mostly concerned with electromagnetics with respect to the pseudo group P_4 or any appropriate subgroup of P_4.

3. TRANSFORMATION FEATURES OF PHYSICAL FIELDS

A consistent application of covariance, as discussed in the

preceding section, requires a definition of the group of transformations as well as a meaningful definition of transformation behavior of physical fields. We may attempt here to bridge the gap between mathematical definition and the heuristic physical justification for the transformation properties of physical fields. A more rigorous discussion will appear in chapter 2.

The concept of vector fields in space as a rule is intuitively accepted as an essential and necessary abstraction in the discussion of physical problems. The traditional geometric image associated with a vector field is the oriented line element with a magnitude. In some cases the dual picture arises: an oriented surface element with a magnitude.

The two dual vector images are indistinguishable from the point of view of transformation behavior so long as one considers orthogonal Cartesian frames. The situation changes as soon as one cares to consider oblique Cartesian frames. The vectors corresponding to an oriented line element then transform as coordinates of a point whereas vectors with an image corresponding to oriented surface elements transform as the coordinates of a plane.

The distinction in transformation properties just cited is the well known distinction between contravariant and covariant behavior. The mathematical discussion of these concepts within the framework of linear transformations can be found in many competent texts; a particularly useful reference is a classic on higher algebra by M. Bôcher [1907].

A thorough discussion of the aspects of contra- and covariant behavior from a point of view of physics is quite rare, because often it is felt that the distinction just mentioned is not essential for physics. If orthogonal frames are admissible throughout space as suggested by Gauss' topographical experiment (see § 1.2), then it would be very unlikely that this distinction could be of any physical consequence.

However, if a more searching experimental analysis brings out that orthogonal Cartesian frames cannot be extended unlimitedly throughout space then one certainly has to consider the physical implications of this state of affairs and the difference between contra- and co-variant behavior is not any more a purely geometrical distinction. The transition between contra- and co-variant components by means of the fundamental tensor now invokes a physical issue.

This is a statement based on the fact that now the fundamental tensor contains some intrinsic information about the nature of physical space, which is expressed in the non-vanishing components of the Riemann-Christoffel tensor of the metrical field.

For instance, if we accept the general relativistic interpretation of the metric, the components of the fundamental tensor represent gravitational potentials. A transition between contra- and co-variant components of a physical field then is given by a contraction between the field and the gravitational potentials. It would be very inconsistent to suppose that this would not affect the physical meaning of the field. Unfortunately, too many times, it is suggested that this is the case, because spatially this matter may be regarded as a minor issue, magnitude-wise.

Assuming that the issue is not a trivial one, we may point out some typical physical representatives of either vector image in ordinary space. The velocity of a particle clearly has the character of a line element which is called a contravariant vector, whereas the wave vector is a typical representative of a covariant vector. Consecutive planes of equal phase come closer if the magnitude of the wave vector increases. The same is true for the gradient of a potential field: the surfaces of equal potential are closest where the gradient is largest. Hence the magnitude is inversely proportional to the distance of the surfaces, a typical feature of a covariant vector.

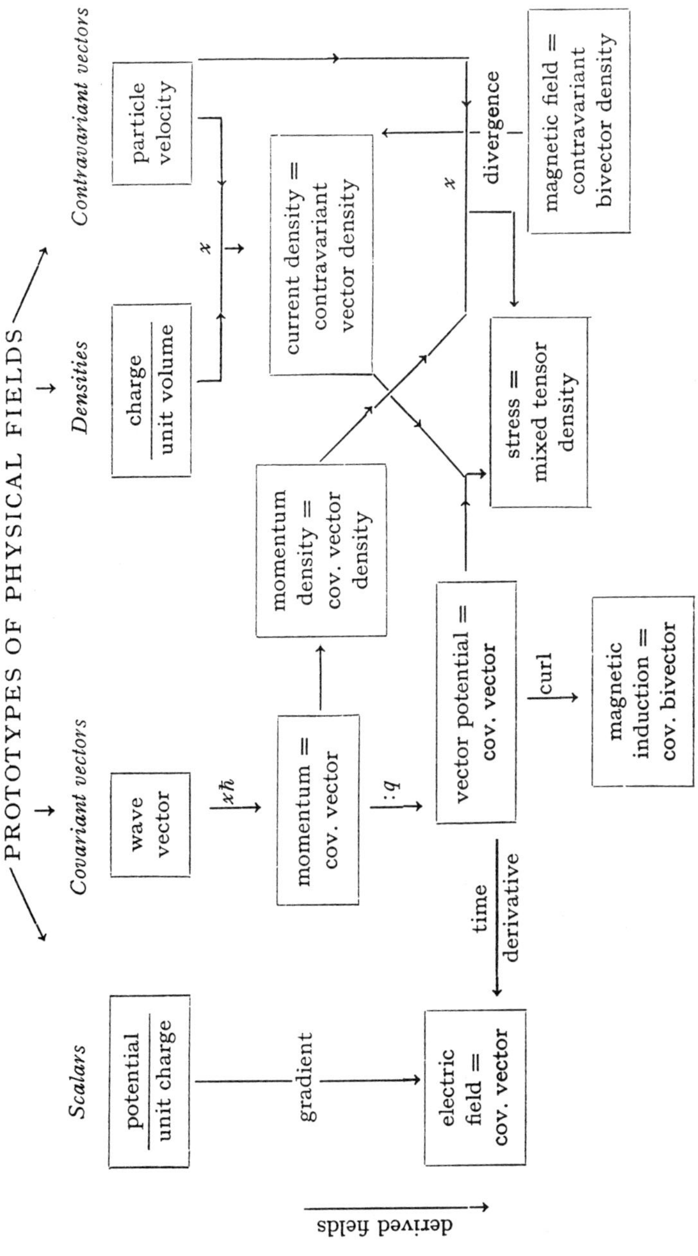

The operations gradient and curl add a covariant valence.

The operation divergence decreases the contravariant valence with one.

$\hbar$ = Planck's quantum of action

q = charge

Fig. 2. Mutual relation of physical fields in space.

Once we have a set of basic fields that can be regarded as prototypes of a specific transformation behavior, we may try some intelligent guesses about the transformation of most of the derived fields. It has been attempted to illustrate this in Fig. 2. The "prototype" fields are given in the blocks of the first row (scalar, covariant vector, density, contravariant vector). A density is a scalar-like quantity that transforms if the unit volume transforms. Following the lines in Fig. 2 it will become clear how the assumed transformation behavior of the "prototype" fields fixes the transformation properties of the derived fields. To convince oneself about the internal consistency it is interesting to note that there is often more than one way leading from the "prototype" to any one specific block in Fig. 2. A brief but exhaustive discussion of this subject matter, including density aspects of physical fields, can be found in a text by J. A. SCHOUTEN [1951].

The merging of space and time covariance, as known, requires a regrouping and completion of spatial fields into four-dimensional fields with physically meaningful transformation properties. This regrouping has to be performed in such a way that for the subgroup of spatial rotations the four-dimensional fields split up again into the constituent spatial fields.

Space-time covariance re-emphasizes the issue of contra- and co-variant behavior, provided one cares to preserve the typical structural properties of the space-time continuum (i.e., no substitutions $x_0 = ict$)†.

The original unadulterated Lorentz metric leads to two versions of a four-dimensional momentum vector

$$(m, \boldsymbol{p}) \quad \text{contravariant}$$
$$(-E, \boldsymbol{p}) \quad \text{covariant},$$

† A. D. FOKKER [1927] has for a long time advocated the consistent use of the original Lorentz metric, leaving the signature of the metric intact.

related by the metric $(-c^2, 1, 1, 1)$, with effective mass m and energy $-E$ as fourth components. The interpretation of the contravariant version breaks down for particles that do not carry a rest mass. The covariant version on the other hand is still meaningful for zero rest mass and can be related to the four-vector $(-\omega, \boldsymbol{k})$ of frequency and wave vector by the invariant factor $\hbar$ (action), thus stressing the intrinsic covariant character of the energy momentum vector (table 1, chapter 2).

The incidental occurrence of such an isolated fact, showing the limitations of the use of the metric for associating co- and contra-variant components can only be regarded as a weak argument in favor of an intrinsic transformation behavior. It is, however, the corroborative nature of the different methods of approach that encourages a general endeavor to investigate the meaning of intrinsic transformation behavior.

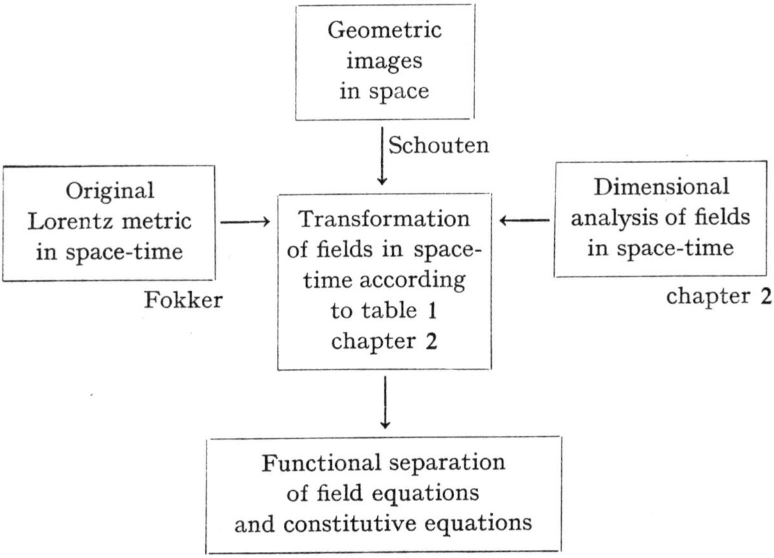

Fig. 3. Comparison of available procedures to investigate transformation properties of physical fields.

The block diagram in Fig. 3 may help the reader as an incentive to check the mutual support of the three approaches to investigate transformation behavior.

The ensuing transformation for the electromagnetic fields will turn out to be the only one that leads to a consistent functional separation of field equations and constitutive equations (chapter 3).

It may be mentioned that Lorentz transformations do not show up density features of the fields, because the Jacobian is one. The method of geometrical images, on the other hand, will show density features in three dimensions in a very convincing manner. The dimensional method integrates and completes the partial conclusions of the two other methods.

4. NOTATIONAL CONVENTIONS

The Absolute Differential Calculus is known as the most suitable mathematical tool when dealing with matters of general covariance. The absolute differential calculus is otherwise known as Ricci Calculus or Tensor Analysis. It should be noted, however, that the name tensor analysis or tensor algebra in many cases covers the more restricted discipline of Cartesian tensors.

The most simple procedure from here on would be to refer to the appropriate standard texts for relevant mathematical information. We feel, however, that this might shy away readers from the actual subject matter. It is true that a training in the recognition of invariant structures may help one to grasp the physical essentials. On the other hand, it should not be forgotten that it is customary to study analytical dynamics without a preknowledge of the absolute differential calculus. Therefore we will attempt a compromise to prevent undue reference work.

Three text books may be recommended as particularly useful for general orientation. They are T. LEVI CIVITA [1926], L. BRILLOUIN [1949] and J. A. SCHOUTEN [1951].

Levi Civita's text is a classic on the subject. As regards the discussion of physical fields it concurs with Einstein's papers in considering only the unimodular (pseudo) subgroup of P_4 (see Fig. 1), thus eliminating density features of physical fields.

Brillouin gives an excellent discussion of the transformation features of physical fields in three dimensions including density properties. As such this work (first edition 1938) for a long time held a unique position in the physical literature, considering the prevalent feelings among physicists about the redundancy of that much transformation detail.

Schouten's text contains the most mathematically up-to-date material, likewise including discussions of density aspects of physical fields. Schouten's notation (kernel-index method) is superior in accentuating the essentials of the absolute differential calculus. A brief exposition of this notation and the algebraic preliminaries are given below. The kernel-index method will be adopted throughout this text.

Transformations

A coordinate transformation will be denoted by a change in label (index) $x^\lambda \rightarrow x^{\lambda'}$, where $x^{\lambda'}$ is an arbitrary function of x^λ,

$$x^{\lambda'} = x^{\lambda'}(x^\lambda)\,, \quad \lambda, \lambda' = 1, 2, \ldots, n \tag{1.1}$$

and λ and λ' label the coordinates in the two different systems†. A point transformation may be denoted by a change in the "kernel" symbol $x^\lambda \rightarrow y^\lambda$. So the sort of index λ, λ' or λ'' characterizes the coordinate system; the "kernel" symbol characterizes the fields.

It will be assumed that the set (1.1) has a unique solution in the domain considered so that

† The use of a same Greek index λ in the primed and in the unprimed system does not imply that λ' represents the same integer as λ.

$$x^{\lambda} = x^{\lambda}(x^{\lambda'})\,. \tag{1.2}$$

The elements of the Jacobian transformation matrices are denoted by the symbols

$$\frac{\partial x^{\lambda'}}{\partial x^{\lambda}} = A^{\lambda'}_{\lambda} \tag{1.3}$$

$$\frac{\partial x^{\lambda}}{\partial x^{\lambda'}} = A^{\lambda}_{\lambda'}\,. \tag{1.4}$$

There is no significance (except convenience) attached to the fact that the same Greek symbol is used in the primed and the unprimed systems. We could just as well use $A^{\lambda'}_{\sigma}$ or $A^{\kappa}_{\sigma'}$ etc.

There is no danger of confusing transformation elements $A^{\lambda'}_{\lambda}$ with tensors because they always carry indices of a different kind, λ' and λ belonging to different frames of reference.

If in a transformation formula a product of $A^{\lambda'}_{\lambda}$ occurs, then the "kernel" symbol A may not be repeated as it is irrelevant for the operations, e.g.,

$$A^{\lambda}_{\lambda'}A^{\sigma}_{\sigma'}A^{\kappa'}_{\kappa} = A^{\lambda\,\sigma\,\kappa'}_{\lambda'\sigma'\kappa}\,. \tag{1.5}$$

The summation convention only applies to a repeated index of which one is in the lower and the other in the upper position (subscript – superscript summation only), e.g.,

$$\sum_{\lambda} A^{\lambda}_{\sigma'}A^{\kappa'}_{\lambda} = A^{\lambda}_{\sigma'}A^{\kappa'}_{\lambda}\,. \tag{1.6}$$

Using the convention (1.5), this can be written

$$\sum_{\lambda} A^{\lambda}_{\sigma'}A^{\kappa'}_{\lambda} = A^{\lambda\,\kappa'}_{\sigma'\lambda}\,. \tag{1.7}$$

Other summations are not meaningful in a general covariant

sense. The summation does not apply to indices of a different kind λ', λ, or ν', ν, or indices that are both subscripts or superscripts.

Using the present notation Jacobi's law for two consecutive transformations reads

$$A^{\lambda}_{\lambda'}A^{\lambda'}_{\lambda''} = A^{\lambda\,\lambda'}_{\lambda'\lambda''} = A^{\lambda}_{\lambda''}\,, \tag{1.8}$$

and the fact that (1.3) and (1.4) are reciprocal is expressed by

$$A^{\lambda\,\sigma'}_{\sigma'\nu} = A^{\lambda}_{\nu} = \delta^{\lambda}_{\nu} \quad \text{or} \quad A^{\lambda\,\nu'}_{\kappa'\lambda} = A^{\nu'}_{\kappa'} = \delta^{\nu'}_{\kappa'}\,; \tag{1.9}$$

A^{λ}_{ν} in fact is the unit tensor identical to Kronecker's delta. It has the same components in every system of reference (1 on the main diagonal). It is to be noted that this is an invariant feature only if the transformation behavior is mixed; i.e., one label covariant, the other contravariant.

The determinant of the Jacobian matrix is usually denoted by Δ:

$$|A^{\lambda'}_{\lambda}| = \Delta\,. \tag{1.10}$$

The reciprocal transformation elements are related by

$$A^{\lambda}_{\lambda'} = \frac{\partial \ln \Delta}{\partial A^{\lambda'}_{\lambda}}\,. \tag{1.11}$$

Differentiation with respect to the coordinates is denoted by

$$\partial_{\nu} = \frac{\partial}{\partial x^{\nu}}\,, \qquad \partial_{\nu'} = \frac{\partial}{\partial x^{\nu'}}\,, \text{ etc.} \tag{1.12}$$

Holonomic transformation elements are derivable from integral relations (1.1) and (1.2). They satisfy (if well behaved)

$$\partial_\nu A^{\lambda'}_\lambda = \partial_\lambda A^{\lambda'}_\nu \quad \text{and} \quad \partial_{\nu'} A^{\lambda}_{\lambda'} = \partial_{\lambda'} A^{\lambda}_{\nu'} \,. \tag{1.13}$$

The transformation is called anholonomic if (1.13) is not satisfied which means that integral relations of the kind (1.1) and (1.2) do not exist. The coefficients $A^{\lambda'}_\lambda$ are then arbitrary functions of the coordinates, they are not defined any more by (1.3) and (1.4).

Geometrical Objects

Mathematical constructs with linear transformation behavior occurring in geometry and physics are called "objects". The most important subclasses are called quantities, tensors or affinors. They obey a linear homogeneous law of transformation, and are the natural generalizations of the Cartesian vectors and tensors, e.g.,

$$B^{\sigma'}_{\cdot\kappa'\nu'} = |\Delta|^{-k} A^{\sigma'\kappa\,\nu}_{\sigma\,\kappa'\nu'} B^{\sigma}_{\cdot\kappa\nu} \,. \tag{1.14}$$

The terminology is that κ, ν are covariant labels, σ is a contravariant label, depending on whether they are associated with the reciprocal $A^{\kappa}_{\kappa'}$ or direct $A^{\sigma'}_{\sigma}$ transformation elements. The tensor $B^{\sigma}_{\cdot\kappa\nu}$ is of mixed valence three, covariant valence two, contravariant valence one. The exponent k gives the weight of the transformation of $B^{\sigma}_{\cdot\kappa\nu}$. Quantities or tensors with a weighted transformation (i.e., $k \neq 0$) are called densities and are usually denoted by a Gothic or Greek "kernel" symbol. Brillouin has coined the names "intensities" and "capacities" for the densities of weight $+1$ and -1; they occur most often in physics†.

The transformation of an "object" is as a rule not homogeneous and may invoke the derivatives of the transformation elements. An important example in relativity is the Christoffel

† Please compare with chapter 2, § 6 for the distinction between densities and pseudo densities.

symbol. On holonomic coordinates it is

$$\Gamma^{\lambda}_{\nu\kappa} = \tfrac{1}{2}(\partial_\nu g_{\sigma\kappa} + \partial_\kappa g_{\sigma\nu} - \partial_\sigma g_{\nu\kappa})g^{\sigma\lambda}, \tag{1.15}$$

in which $g_{\nu\kappa}$ and $g^{\sigma\lambda}$ are the metric tensor and its reciprocal. It obeys the transformation rule

$$\Gamma^{\lambda'}_{\nu'\kappa'} = A^{\lambda'\nu\kappa}_{\lambda\,\nu'\kappa'}\Gamma^{\lambda}_{\nu\kappa} + A^{\lambda'}_{\sigma}\partial_{\nu'}A^{\sigma}_{\kappa'}\,, \tag{1.16}$$

which incidentally holds for holonomic and anholonomic transformations.

It is to be noted that the transformation of this "object" reduces to the transformation of a tensor for the group of linear transformations. The important difference with a tensor is that $\Gamma^{\lambda'}_{\nu'\kappa'}$ can be different from zero even if $\Gamma^{\lambda}_{\nu\kappa}$ vanishes. It should be noted that this is impossible for (1.14). The components of $\Gamma^{\lambda'}_{\nu'\kappa'}$ depend on the rate of change of the transformation elements $\partial_{\nu'}A^{\sigma}_{\kappa'}$, hence in space-time it depends on the rate of change with time. One may therefore expect $\Gamma^{\lambda}_{\nu\kappa}$ to occur in association with inertia forces, because an inertia force can never be a vector in the general relativistic sense.

Invariant Operations

Invariant algebraic processes are addition, multiplication, contraction, alternation and mixing. The last three lead to typical notational conventions. *Contraction* is summation over a lower and upper index as already mentioned for the transformation elements. Alternation and mixing operations only apply to either contra- or co-variant indices, otherwise there is no invariant meaning.

Alternation is a determinant-like operation. It is denoted by square brackets around the indices and defined by the formula

$$P_{[\kappa\lambda]} = \frac{1}{2!}(P_{\kappa\lambda} - P_{\lambda\kappa})\,, \tag{1.17}$$

$$P_{[\kappa\lambda\nu]} = \frac{1}{3!}(P_{\kappa\lambda\nu} + P_{\nu\kappa\lambda} + P_{\lambda\nu\kappa} - P_{\kappa\nu\lambda} - P_{\lambda\kappa\nu} - P_{\nu\lambda\kappa})\,,\ \text{etc.}\,,$$

even permutations with + sign, odd permutations with − sign.

Mixing is an operation similar to alternation except that the odd permutations get a + sign too. It is denoted by putting ordinary brackets around the relevant indices

$$P_{(\nu\lambda)} = \tfrac{1}{2}(P_{\nu\lambda} + P_{\lambda\nu}) \text{ etc.} \tag{1.18}$$

Any n-dimensional manifold has always three fields with numerically invariant components for any transformation. These fields exist *a priori*, independent of the structure (metric or linear displacement) of the manifold and are very useful for invariant algebraic manipulations. They are (normalized to unity): the unit tensor

$$A_\nu^\lambda = \begin{cases} 1 & \lambda = \nu \\ 0 & \lambda \neq \nu \text{ (see (1.9))} \end{cases} \tag{1.19}$$

and the two alternating tensor densities of valence n

$$\left.\begin{array}{l} \underset{\text{weight }+1}{\mathfrak{E}^{1\ldots n} = \mathfrak{E}^{[1\ldots n]}} \\ \\ \underset{\text{weight }-1}{\mathfrak{e}_{1\ldots n} = \mathfrak{e}_{[1\ldots n]}} \end{array}\right\} = \left\{\begin{array}{ll} +1 \text{ for even permutations of } 1 \ldots n & (1.20) \\ \\ -1 \text{ for odd permutations of } 1 \ldots n\,. & (1.21) \end{array}\right.$$

The fields (1.20) and (1.21) are called the permutation fields or the permutation densities of a manifold of n dimensions. The proof of the numerical invariance is simple, the Jacobian drops out if one carries out the transformation, except for the sign of $|\Delta|/\Delta$. The invariance, therefore, holds only for proper transformations ($\Delta > 0$). One obtains an invariance with respect to proper ($\Delta > 0$) and improper ($\Delta < 0$) transformations if one postulates that the permutation field should transform with the Jacobian itself instead of with its absolute value as in (1.14) (see § 2.6, pseudo tensors).

It is only through the progress of science in recent times that we have become acquainted with so large a number of physical quantities that a classification of them is desirable.

J. C. MAXWELL
Proc. London Math. Soc.
Vol. III (1871) 224

CHAPTER 2

TRANSFORMATION BEHAVIOR AND DIMENSIONAL PROPERTIES OF PHYSICAL FIELDS

1. GENERAL

DORGELO and SCHOUTEN [1946] and SCHOUTEN [1951] have associated in a straightforward way the principle of covariance with the concept of physical dimensions. Their considerations led them to distinguish an invariant factor in the dimensional characterization of physical quantities, called the absolute dimension. The actual dimensions of the components of a physical quantity then are determined by this absolute dimension times a multiplicative contribution solely determined by the transformation behavior of the physical quantity.

Dorgelo and Schouten apply these principles in three-space for a discussion of the MKS system. In their discussion of covariance in four-space the authors seem to adhere to the geometrist's viewpoint of assuming dimensional homogeneity of the space-time manifold.

Our present aim is to show that one arrives at a rather striking disentanglement of the dimensional picture if one drops the tacit assumption of dimensional homogeneity of the coordinates of the space-time manifold.

The dimensional distinction between position and time coordinates has been regarded as a disturbing asymmetry ever since the "geometry" of four-space started to play a role in physics.

However, the common dimensional identification of position and time coordinates is acceptable only as an expedient. It facilitates the exploration of the basic physical relations in four-dimensional garb, but it is unacceptable as a lasting feature, because it eliminates the undeniable difference in dimensional individuality of the time and space coordinates. The tacit assumption of dimensional homogeneity is bound to lead to obstacles, because it ignores a fundamental principle, the principle of causality.

Hence rather than blotting out the difference between time and space coordinates we will assume from here on that the observer, whatever his state of motion, is always capable of distinguishing between the time and spatial properties of the manifold. We shall call this the principle of dimensional coordinate individuality in the space-time continuum. A set of coordinates that does not permit this time-space distinction will be denoted as a set of coordinates without an immediate physical meaning for the observer. In the set of general space-time transformations we single out the subset that preserves the physical characteristics of space and time coordinates in the sense of the aforementioned principle of dimensional coordinate individuality. This subset is still a set with non-linear transformations in space-time and will consequently still invoke the principle of general covariance.

The question arises whether a structure can be erected satisfying requirements of space-time covariance and co-ordinate individuality.

2. THE MODIFIED VEBLEN-WHITEHEAD MODEL

VEBLEN and WHITEHEAD [1932], in a booklet entitled Foundations of Differential Geometry, have developed an analytic model which seems to accommodate most branches of differential geometry that came to life before and after the era of relativity.

The name "geometry" for this field of abstract analysis, quoting the authors, is a choice on emotional and traditional grounds depending on the number of competent people that believe it to be right.

In the light of the principle of coordinate individuality a slightly more objective definition seems possible, e.g., the analysis of a manifold that is characterized by dimensional coordinate-homogeneity may be called a geometry.

Physics as the analysis of human perceptive experience in the realms of space-time is not a geometry in the sense of this definition.

It is necessary of course to distinguish between essential and nonessential dimensional individuality of coordinates. A transition from Cartesian to polar coordinates in ordinary space, for instance, introduces a nonessential dimensional distinction.

An essential dimensional distinction cannot be removed without affecting the objective meaning of the description. The essential nature of a dimensional distinction between time and position coordinates shall be accepted as an empirical fact†.

We may proceed to sketch a modified Veblen-Whitehead model to accommodate dimensional individuality of coordinates. It should be understood that this model is a design for a housing to accommodate descriptions of physical observations. As a preliminary design one of its first requirements should be that it is a dimensionally fitting accommodation.

It is customary to describe physical observations with respect to "local", linear, space-time frames. These local frames of physical observations constitute a manifold E_{1+3}. This

† A transition $ict \rightarrow x_0$ does affect the objective meaning of the reference frame.

manifold is characterized by the full linear group of real space-time transformations relating the linear frames, with the Lorentz and Galilei groups as important subgroups.

The E_{1+3} has dimensional homogeneity in the three space coordinates only. The time will be regarded as an essentially distinguishable coordinate. Thus the E_{1+3} is not a geometric manifold, but its submanifold E_3 constituted by the three space coordinates is a geometric manifold.

The coordinates of E_{1+3} are x^a, $a = 0, 1, 2, 3$, with dimensional characterization

$$[x^0] = [t] \quad \text{and} \quad [x^1] = [x^2] = [x^3] = [l] , \tag{2.1}$$

where l and t are the symbols to denote length and time dimensions.

Physical observations made with reference to a frame (e.g., an accelerated frame) not contained in the full linear group of the previously mentioned E_{1+3}, define another linear manifold for the recording of physical experience, thus leading to a continuum of E_{1+3} manifolds.

Observations made in different E_{1+3}'s, thus far, are uncorrelated physical events. The correlation of the descriptions associated with different E_{1+3}'s invokes the discussion of the integrability of their respective linear frames. This is conveniently done by the introduction of the artifice of an underlying nonlinear manifold A_{1+3}. The underlying manifold will be chosen as an arithmetic manifold, which means that its coordinates and the fields defined over A_{1+3} are dimensionless. The coordinates of the A_{1+3} shall be denoted by ξ^κ, $\kappa = 0, 1, 2, 3$, with

$$[\xi^\kappa] = [1] , \tag{2.2}$$

where the symbol [1] denotes the numerical nature of ξ^κ. The

A_{1+3} is characterized by the pseudo group of general, real coordinate transformation†.††.

The coordinate relations between the A_{1+3} and the continuum of E_{1+3}'s as a rule are nonholonomic. For infinitesimal increments in ξ^κ one has the relations

$$\mathrm{d}\xi^\kappa = A^\kappa_a x^a , \qquad x^a = A^a_\kappa \, \mathrm{d}\xi^\kappa \tag{2.3}$$

with

$$A^\kappa_a A^a_\lambda = \delta^\kappa_\lambda \quad \text{and} \quad A^\kappa_a A^b_\kappa = \delta^b_a .$$

The A^κ_a and A^a_κ are the connecting quantities between A_{1+3} and the continuum of E_{1+3}'s; they are functions of ξ^κ. The A_{1+3} is said to have an E_{1+3} associated with each of its arithmetic points ξ^κ. The connection is nonholonomic if

$$\partial_{[\nu} A^a_{\kappa]} \neq 0 . \tag{2.4}$$

The relations (2.1), (2.2) and (2.3) lead to the following dimensional characterizations of the connecting quantities

$$\begin{aligned} [A^0_\kappa] &= [t] , \quad [A^1_\kappa] = [A^2_\kappa] = [A^3_\kappa] = [l] \\ [A^\kappa_0] &= [t^{-1}] , \quad [A^\kappa_1] = [A^\kappa_2] = [A^\kappa_3] = [l^{-1}] , \end{aligned} \tag{2.5}$$

where the zero label in the sequence of Latin indices again represents the distinguishable time coordinate in the E_{1+3}'s. The Jacobian determinants of A^a_κ and A^κ_a have the dimensions

$$[\, | A^a_\kappa | \,] = [t l^3] , \qquad [\, | A^\kappa_a | \,] = [t^{-1} l^{-3}] . \tag{2.6}$$

† The A_{1+3} has coordinate homogeneity and might be denoted by A_4.

†† Greek indices are used throughout this text; only in the §§ 2.2, 2.3, 2.4 are they explicitly associated with the coordinates of an arithmetic manifold.

Physical field quantities defined in the E_{1+3}'s give rise to an induced, and by definition dimensionless, field image in A_{1+3}. Conversely, a field in A_{1+3} (dimensionless) induces a physical field in E_{1+3}.

Suppose we have a tensor field $P^{\kappa}_{\cdot\lambda\nu}$ in A_{1+3}, where a tensor denotes the subclass of arithmetic (dimensionless) objects in A_{1+3} that have the common feature of a homogeneous transformation rule. The physical field induced in the E_{1+3}'s then is given by a relation

$$P^{a}_{\cdot bc} = D(P) A^{a}_{\kappa} A^{\lambda}_{b} A^{\nu}_{c} P^{\kappa}_{\cdot\lambda\nu} . \tag{2.7}$$

The coefficient $D(P)$ in the formula (2.7) is the physical gauge factor, characteristic of $P^{a}_{\cdot bc}$†. Formula (2.7) states that the dimensions of the components of $P^{a}_{\cdot bc}$ can be obtained from the relations

$$[P^{a}_{\cdot bc}] = [D]\,[A^{a}_{\kappa}]\,[A^{\lambda}_{b}]\,[A^{\nu}_{c}] , \tag{2.8}$$

because $P^{\kappa}_{\cdot\lambda\nu}$ is purely numerical by definition.

Similarly, for a tensor density, e.g., $\mathfrak{T}_{\lambda}^{\cdot\nu}$ (weight + 1) one has the relation

$$\mathfrak{T}_{b}^{\cdot a} = D(\mathfrak{T})\,|\,A^{a}_{\kappa}\,|^{-1}\,A^{a}_{\nu} A^{\lambda}_{b} \mathfrak{T}_{\lambda}^{\cdot\nu} , \tag{2.9}$$

where $D(\mathfrak{T})$ again represents a physical gauge factor characteristic for $\mathfrak{T}_{b}^{\cdot a}$, leading to the dimensional relation

$$[\mathfrak{T}_{b}^{\cdot a}] = [D(\mathfrak{T})]\,[\,|\,A^{a}_{\kappa}\,|^{-1}]\,[A^{a}_{\nu}]\,[A^{\lambda}_{b}] . \tag{2.10}$$

† The coefficients A^{a}_{κ} etc. only take care of the length and time dimensions in the transition $A_{1+3} \to E_{1+3}$. The gauge factor in (2.7) is required to introduce physical dimensions not associated with length and time (see § 2.3).

Following the terminology of Dorgelo and Schouten, the dimension of the gauge factor $[D]$ is called the absolute dimension of the field.

The dimensions of the components of the field are the relative (or common) dimensions with respect to a particular space-time frame in E_{1+3}. They differ whether a, b and c are time labels or space labels, or if a coordinate system is admitted in E_3 with a (nonessential) dimensional inhomogeneity.

In words the contents of (2.8) and (2.10) can be expressed as follows from (2.5) and (2.6):

For a linear frame in E_{1+3} the (relative) dimensions of the components of a tensor field can be found by adding to the absolute dimension a factor l or t for every contravariant index, a factor l^{-1} or t^{-1} for every covariant index, depending on whether the index is a space label or a time label. A density of weight $+k$ obtains in addition a common factor $t^{-k}l^{-3k}$.

3. THE DIMENSIONAL DECOMPOSITION OF PHYSICAL FIELDS

The rules laid down in the previous section give all the essentials for an invariant dimensional decomposition of physical fields in E_{1+3}, provided we stipulate some necessary requirements about the gauge factor D.

The dimension of the gauge factor D should be expressible in the dimensions of invariant physical units only.

The only units which are scalars in space-time that can be formed out of the common quadruple: mass m, charge q, length l and time t are

$$\begin{gathered}\text{the unit of action } [\hbar] = [ml^2t^{-1}]\,, \\ \text{the unit of charge } [q]\,;\end{gathered} \tag{2.11}$$

m, l and t individually are not admissible as constituent factors in $[D]$.

The general invariance of $\hbar$ and q is a known fact that can be accepted on intuitive grounds because both of them occur in a quantized manner and consequently are in principle countable entities.

If one imposes the condition that the absolute dimension $[D]$ *of physical tensor fields in* E_{1+3} *is expressible in* $[\hbar]$ *and* $[q]$ *only, then the transformation behavior of the fields is determined unambiguously by the dimensional relations of the kind* (2.8) *and* (2.10), *if the dimensions of the components of the fields are to check with the customary relative dimensions expressed in* m, q, l, *and* t.

A duality with respect to the always existing permutation fields (1.20) and (1.21) does not affect this statement, because their intrinsic numerical features, in A_{1+3} as well as in E_{1+3}, are easily confirmed by means of (2.5) and (2.6)†.

It should be understood that the uniqueness of this relation between dimensional decomposition and transformation behavior is an empirical observation.

The general procedure now is (for instance for the energy momentum tensor) that one compares the different possibilities: double covariant, double contravariant, mixed, density weight, etc., until one has a gauge factor D expressed in $\hbar$ and q only. To cut a long story short we may show the tabulated results for some of the most important physical fields in table 1. The table should give a rather self-explanatory account of the invariant dimensional decomposition illustrated by the expressions (2.8) and (2.10). The choice in table 1, associated with duality, has been decided on grounds of simplicity and tradition†.

In any properly invariant physical relation the coordinate

† For instance the current density $\mathfrak{c}^d$ leads to a dual structure c_{abc}, a covariant trivector alternating in abc. The dimensions associated with the transformation of $\mathfrak{c}^d$ are $\rightarrow l^{-3}t^{-1}l^{+1}q$ and for $c_{abc} \rightarrow l^{-2}t^{-1}q$ etc.

dimensions take care of themselves. This observation leads to a useful corollary of the rules of dimensional decomposition.

In a covariant physical relation the dimensions of the gauge factors of the constituent fields are individually compatible from a dimensional point of view.

As an example we take the linear relations between the electromagnetic fields,

$$\mathfrak{G}^{\lambda\nu} = \tfrac{1}{2}\chi^{\lambda\nu\sigma\kappa}F_{\sigma\kappa}\,. \tag{2.12}$$

The dimensions of the gauge factors of the two fields can be obtained from table 1. The compatibility of gauge dimensions (absolute dimension) in (2.12) then leads to

$$[D(\chi)] = \left[\frac{D(\mathfrak{G})}{D(F)}\right] = [q^2\hbar^{-1}]\,. \tag{2.13}$$

In (2.12) $\mathfrak{G}^{\lambda\nu}$ represents a density of weight $+1$, $F_{\sigma\kappa}$ is a nondensity (weight 0), hence $\chi^{\lambda\nu\sigma\kappa}$ should be a tensor density of weight $+1$. Consequently, the dimensional decomposition of χ in E_{1+3} yields

$$[\chi^{abcd}] = \left[\begin{array}{c|c} \chi^{0i0k} & \chi^{hi0k} \\ \hline \chi^{0ijk} & \chi^{hijk} \end{array}\right]_{h,\,i,\,j,\,k=1,\,2,\,3}$$

$$= [q^2\hbar^{-1}]\,[t^{-1}l^{-3}]\left[\begin{array}{c|c} t^2l^2 & tl^3 \\ \hline tl^3 & l^4 \end{array}\right] = \left[\begin{array}{c|c} q^2\hbar^{-1}tl^{-1} & q^2\hbar^{-1} \\ \hline q^2\hbar^{-1} & q^2\hbar^{-1}t^{-1}l \end{array}\right]$$

$$= \left[\begin{array}{c|c} \text{dielectric matrix} & \tilde{\gamma} \\ \hline \tilde{\gamma} & \text{inverse permeability matrix} \end{array}\right]. \tag{2.14}$$

TABLE 1

Dimensional decomposition of physical tensor fields in space-time†

Physical tensor fields and their transformation in E_{1+3}		Dimension associated with the physical gauge (absolute dimension)	Dimension associated with the weighting factor	Dimension associated with time and space coordinates	Identification in three-space	
four-vector of energy momentum	p_a	$[\hbar]$	$[1]$	$\begin{bmatrix} t^{-1} \\ l^{-1} \end{bmatrix}$	$[ml^2t^{-2}]$ $[mlt^{-1}]$	energy momentum
four-potential of electromagnetics	A_a	$[q^{-1}\hbar]$	$[1]$	$\begin{bmatrix} t^{-1} \\ l^{-1} \end{bmatrix}$	$[q^{-1}ml^2t^{-2}]$ $[q^{-1}mlt^{-1}]$	scalar potential vector potential
four-vector of particle stream	$\mathfrak{N}^a$	$[1]$	$[l^{-3}t^{-1}]$	$\begin{bmatrix} t \\ l \end{bmatrix}$	$[l^{-3}]$ $[l^{-2}t^{-1}]$	particle density stream density
four-vector of current density	$\mathfrak{C}^a$	$[q]$	$[l^{-3}t^{-1}]$	$\begin{bmatrix} t \\ l \end{bmatrix}$	$[ql^{-3}]$ $[ql^{-2}t^{-1}]$	charge density current density

four-vector of force density	$\mathfrak{F}_a$	$[\hbar]$	$[l^{-3}t^{-1}]$	$\begin{bmatrix} t^{-1} \\ l^{-1} \end{bmatrix}$	$[ml^{-1}t^{-3}]$ $[ml^{-2}t^{-2}]$	power density force density
six-vector density of displacement and magnetic field	$\mathfrak{G}^{ab}$	$[q]$	$[l^{-3}t^{-1}]$	$\begin{bmatrix} lt \\ l^2 \end{bmatrix}$	$[ql^{-2}]$ $[ql^{-1}t^{-1}]$	dielectric displacement $\boldsymbol{D}$ magnetic field $\boldsymbol{H}$
six-vector of electric field and magnetic induction	F_{ab}	$[q^{-1}\hbar]$	$[1]$	$\begin{bmatrix} l^{-1}t^{-1} \\ l^{-2} \end{bmatrix}$	$[q^{-1}mlt^{-2}]$ $[q^{-1}mt^{-1}]$	electric field $\boldsymbol{E}$ magnetic induction $\boldsymbol{B}$
energy and momentum tensor	$\mathfrak{T}_a{}^b$	$[\hbar]$	$[l^{-3}t^{-1}]$	$\left[\begin{array}{c\|c} 1 & lt^{-1} \\ \hline l^{-1}t & 1 \end{array}\right]$	$= \left[\begin{array}{c\|c} ml^{-1}t^{-2} & mt^{-3} \\ \hline ml^{-2}t^{-1} & ml^{-1}t^{-2} \end{array}\right]$ $= \left[\begin{array}{c\|c} \text{energy density} & \text{energy flux} \\ \hline \text{momentum density} & \text{stress} \end{array}\right]$	

† Sub- and superscripts denote co- and contravariant behavior, gothic kernel symbol denotes density of weight + 1; $\hbar = ml^2t^{-1}$ = action, q = charge, l = length, t = time, m = mass, [1] denotes a dimensionless factor.

The terms in the off-diagonal squares of (2.14) denoted by the matrix $\tilde{\gamma}$ are associated with the Fresnel-Fizeau effect and the natural optical activity (chapter 8).

4. THE FUNDAMENTAL TENSOR

An interesting point arises if we ask the question whether the fundamental tensor should be treated in the same manner as the other physical fields, because the fundamental tensor is closely connected with the choice of units for length and time. It is customary to associate the line element in space-time with a length dimension; however, a time dimension could have been chosen equally well. A line element with the dimension of a length or time violates our previously adopted rule that the dimension of a scalar field can depend only on the dimensions of the invariant units of action $[\hbar]$ and of charge $[q]$. The line element with a length dimension implies that the fundamental tensor g_{ab} should carry an absolute dimension of $[l^2]$ as stipulated by Dorgelo and Schouten. The reciprocal g^{ab} of the fundamental tensor then has an absolute dimension $[l^{-2}]$. An acceptable alternative in agreement with our present system would be to introduce a normalized fundamental tensor with a dimensionless gauge factor. This normalized fundamental tensor is obtained from the customary fundamental tensor by the relation

$$\text{normalized } g_{ab} = l^{-2} \text{ customary } g_{ab}\,, \tag{2.15}$$

where l is a definite physical length. The elements of the normalized fundamental tensor then have dimensions according to the position of the indices as denoted in (2.16)

$$g_{ab} = \left[\begin{array}{c|c} t^{-2} & l^{-1}t^{-1} \\ \hline l^{-1}t^{-1} & l^{-2} \end{array}\right], \quad g^{ab} = \left[\begin{array}{c|c} t^{2} & lt \\ \hline lt & l^{2} \end{array}\right], \tag{2.16}$$

and the square root of the determinant of the metric, which

is known to play the role of a density, obtains a dimension in accordance with the characteristics of a density of weight + 1, i.e.,

$$[g^{\frac{1}{2}}] = [\,|\,g_{ab}\,|^{\frac{1}{2}}\,] = [l^{-3}t^{-1}]\,. \tag{2.17}$$

Another feature of interest for the normalized metric is that the dimension of the magnitude of a field is equal to the dimension of its gauge factor or absolute dimension. This point is illustrated by formula (2.18) for the six-vector of the electric field and magnetic induction

$$D(F_{ab}) = [F_{ab}F_{cd}g^{ac}g^{bd}]^{\frac{1}{2}} = [q^{-1}\hbar]\,. \tag{2.18}$$

For our present purpose we shall find in chapter 9 that the question whether or not a gauge factor for g_{ab} should be used is not too urgent for electromagnetism. Either the gauge factor of the metric or the metric itself cancels out in all quantities and equations of importance. For the Maxwell equations this is true when we consider that they can be formulated independently of the metric. It will be shown, however, in chapter 9 that the constitutive tensor of matter-free space, though dependent on the metric, is still independent of the choice of the gauge factor of the metric because g_{ab} and g^{ab} occur homogeneously to the same degree.

In conclusion it may be of some interest to note that the normalization factor mentioned in (2.15) is reminiscent of WEYL's [1921] calibration factor for the metric that is associated with a nonintegrability of the space-time interval. It seems that a normalized metric in a sense supports Weyl's question whether a Riemann geometry with nonintegrability of direction only is adequate for a complete description of physical phenomena. The validity of Weyl's criticism, however, is not necessarily a support for his unification of gravitation and electromagnetism. As mentioned before, we do not have to commit ourselves with the physical implications of a normalization factor of the metric so long as we stay within the realms of electromagnetism.

5. NONTENSORIAL PHYSICAL FIELDS

A most fundamental nontensorial field that has been considered in physics is the field of linear displacement Γ^c_{ab} occurring in the equation of the geodesic lines. The transformation behavior of this field is very similar to that of a tensor field except that an inhomogeneous term must be added. The relation between Γ^c_{ab} and its arithmetic image in A_{1+3} is known to be

$$\Gamma^c_{ab} = D(\Gamma)\,\{A^c_\kappa A^\lambda_a A^\nu_b \Gamma^\kappa_{\lambda\nu} - A^\lambda_a A^\nu_b \partial_\lambda A^c_\nu\} \tag{2.19}$$

(please compare with (2.7)). Relation (2.19) is valid for holonomic as well as nonholonomic frames (see (2.4)).

As in (2.7), a physical gauge factor $D(\Gamma)$ has been formally inserted in (2.19). The dimensional consistency of the expressions for the covariant differential however, e.g.,

$$\delta v^a = \mathrm{d}v^a + \Gamma^a_{bc} v^b \mathrm{d}x^c\,, \tag{2.20}$$

requires that $D(\Gamma)$ should be purely numerical, i.e.,

$$[D(\Gamma)] = [1]\,. \tag{2.21}$$

Recalling the, by definition, numerical nature of the field $\Gamma^\kappa_{\lambda\nu}$ in A_{1+3}, we find that equation (2.19) with (2.5) and (2.21) yields the dimensions of Γ^a_{bc} in E_{1+3}. In other words, the rule formulated at the end of § 2.1 applies, i.e.,

Γ^a_{bc}	oo	os	sr
o	t^{-1}	l^{-1}	tl^{-2}
r	lt^{-2}	t^{-1}	l^{-1}

(2.22)

o = time coordinate; $s, r = 1, 2, 3$ are the space coordinates.

The components of Γ^a_{bc} have tensor properties within the confines of the full linear group in E_{1+3} because the transformation elements of the linear group are constants so that the nonhomogeneous term in (2.19) vanishes.

It is possible to accommodate classical descriptions in the space-time formalism if one disassociates the Γ^a_{bc} from the fundamental tensor. The field of linear displacement then disintegrates in some familiar physical fields if one maintains absolute time. One may consider for instance the simple case of a transition from an inertial frame to a rotating frame. The inertial frame being denoted by labels a', b', c', the rotating frame by the labels a, b, c, the inertial frame is characterized by

$$\Gamma^{a'}_{b'c'} = 0\,. \tag{2.23}$$

The rotation of the $a\,b\,c$ frame then generates a field Γ^a_{bc}, which according to (2.19) is given by

$$\Gamma^a_{bc} = -\,A^{b'}_b A^{c'}_c \partial_{b'} A^a_{c'}\,. \tag{2.24}$$

One finds after some calculations:

$$\Gamma^r_{00} = \text{the vector field of centrifugal acceleration;} \tag{2.25}$$

$$\Gamma^r_{0s} = \text{the bivector of angular velocity } (r, s = 1, 2, 3)\text{;} \tag{2.26}$$

and the other coefficients vanish. On an orthogonal space frame, $\Gamma^r_{0s} = \Gamma^s_{0r}$, a familiar property of Ricci's coefficients of rotation. The coefficients Γ^a_{bc} in this context, though, represent the kinematical information that relates the rotating frame to an inertial frame. A comparison of (2.25) and (2.26) with (2.22) confirms the dimensional decomposition obtained from (2.19) and (2.21) as a physically meaningful one. Forces are an important class of physical fields closely associated with the Γ^a_{bc}. They are in general nontensorial because they can be created by a change in the frame of reference. However, the sum of inertial and gravitational forces may be tensorial. The Lorentz force is another example of a tensorial force field if it is regarded as a contravariant vector density according to table 1.

The interpretation of the coefficients Γ^a_{bc} is much more complex if the linear displacement is associated with a fundamental tensor in the space-time manifold. It should be noted that the Γ^a_{bc} still do not vanish though the g_{ab} are constants on the linear frames in E_{1+3}. The following expression may be extracted from the literature (SCHOUTEN [1954], p. 170) for the mutual relation between the linear displacement and the fundamental tensor on linear local coordinates

$$\Gamma^a_{bc} = \tfrac{1}{2}g^{ad}\{\Omega_{bcd} - \Omega_{dbc} + \Omega_{cdb}\}, \tag{2.27}$$

with

$$\Omega_{bcd} = g_{ed}A^\lambda_b A^\nu_c \partial_{[\lambda}A^e_{\nu]}. \tag{2.28}$$

The reader may convince himself that (2.27) checks dimensionally with (2.22) independent of the assumption (2.15) about the gauge factor of the fundamental tensor.

The coefficients of linear displacement are the constituent elements of the Riemann-Christoffel tensor of the manifold. The R.C. tensor, the contracted R.C. tensor, as well as the Einstein tensor density that occurs in the gravitational field equations in E_{1+3}, all have a dimensionless gauge factor because the field of linear displacement and the fundamental tensor have dimensionless gauge factors [(2.15) and (2.21)]. The gravitational scalar (constant) that relates the Einstein tensor to the energy momentum tensor, should therefore have the dimension $[\hbar^{-1}]$ if the field equations are covariant equations (see § 2.2). For the so-called "scalar curvature" we arrive at the conclusion that it should be dimensionless.

A dimensionless "scalar curvature" is the only logically acceptable possibility for this quantity because there is no empirical evidence whatever that the space-time manifold is embedded in a higher dimensional manifold having a metric whose units can be used to measure the radius of curvature of the space-time manifold.

6. PSEUDO TENSORS

A class of physical quantities which has not yet been discussed are the pseudo tensors. Pseudo tensors, like tensors, follow homogeneous transformation laws. The distinction with tensors lies in the behavior of the weighting function with respect to improper coordinate transformations ($\Delta < 0$). The weighting function of the pseudo tensor changes its sign for an improper transformation. The ordinary tensor on the other hand has either no weighting function (unity) or it transforms with the absolute value of the Jacobian (see e.g., (1.14)). In the latter case the ordinary tensor is called a tensor density, or ordinary tensor density to stress the difference with the so-called pseudo-tensor densities. The ordinary tensor in this connection may be regarded as a density of weight zero. It will be clear that the $\pm$ sign of the weighting factor does not affect the physical dimensions; therefore, transformation detail associated with the $\pm$ sign of the weighting function has to be investigated by other than dimensional means.

We may attempt first to delineate a little bit more precisely the mathematical differences between tensors and pseudo tensors in order to check in what respect the here mentioned distinction may be regarded as really unique. In § 1.4 we defined the tensor densities (1.14) and the so-called permutation fields (1.20) and (1.21). The latter are densities of weight $+1$ and -1. The weighting factor in all three cases was defined with the absolute value of the Jacobian determinant $|\Delta|$, really for no other reason than that this leads to a most useful and frequently occurring representation of physical fields. Hence, removing this limitation, we may now consider fields which transform with the Jacobian itself instead of with its absolute value only. Following Schouten we shall denote the pseudo-tensor fields by a "tilde" over the kernel symbol. A contravariant pseudo vector then transforms according to the rule

$$\tilde{V}^{\lambda'} = \frac{\Delta}{|\Delta|} \tilde{V}^{\lambda} A_{\lambda}^{\lambda'} . \tag{2.29}$$

The $\pm$ sign. sensitivity for improper transformations is in the factor $\Delta / |\Delta|$, which always has the absolute value unity.

A covariant pseudo vector obeys the rule

$$\tilde{W}_{\lambda'} = \frac{\Delta}{|\Delta|} \tilde{W}_{\lambda} A_{\lambda'}^{\lambda} , \tag{2.30}$$

and a pseudo-vector density of weight + 1 obeys

$$\tilde{\mathfrak{B}}^{\lambda'} = \Delta^{-1} A_{\lambda}^{\lambda'} \tilde{\mathfrak{B}}^{\lambda} . \tag{2.31}$$

The "pseudo" features of course disappear for a density of even weight unless one cares to add in the factor $\Delta / |\Delta|$.

It was already mentioned in the last part of § 1.4 that the permutation fields (1.20) and (1.21) can be converted into pseudo densities. Their transformation rules for a three-dimensional manifold are then

$$\tilde{\mathfrak{E}}^{\lambda'\nu'\sigma'} = \Delta^{-1} A_{\lambda\ \nu\ \sigma}^{\lambda'\nu'\sigma'} \tilde{\mathfrak{E}}^{\lambda\nu\sigma} \tag{2.32}$$

$$\tilde{\mathfrak{e}}_{\lambda'\nu'\sigma'} = \Delta A_{\lambda'\nu'\sigma'}^{\lambda\ \nu\ \sigma} \tilde{\mathfrak{e}}_{\lambda\nu\sigma} . \tag{2.33}$$

The components of the fields (2.32) and (2.33), in contradistinction to (1.20) and (1.21), are absolute invariants with respect to proper as well as improper transformations. The (normalized) pseudo permutation fields are, therefore, very useful algebraic operators for deriving a so-called dual equivalent of any tensor, or pseudo tensor, field. This dual equivalence is then valid for proper and improper transformations, whereas the duals obtained by (1.20) and (1.21) are only meaningful for proper transformations.

It will be clear from the formulae (2.29) . . . (2.31) that the following rule holds for the direct or contracted multiplications of tensors and pseudo tensors: the product of two pseudo tensors yields a tensor, the product of a tensor and a pseudo tensor is a pseudo tensor.

We may now discuss a physical example in three dimensions. Suppose that the magnetic field in three dimensions should be considered as a skew symmetric contravariant tensor density with transformation†

$$\mathfrak{H}^{\lambda'\nu'} = |\Delta|^{-1} A^{\lambda'\nu'}_{\lambda\ \nu} \mathfrak{H}^{\lambda\nu}, \qquad \lambda, \nu = 1, 2, 3\,. \tag{2.34}$$

A contraction over λ and ν with the (pseudo) permutation field (2.33) then yields as the dual of (2.34) the pseudo vector

$$\tilde{H}_\sigma = \tfrac{1}{2}\tilde{\mathfrak{e}}_{\lambda\nu\sigma}\mathfrak{H}^{\lambda\nu}. \tag{2.35}$$

In other words, the magnetic field, considered as a vector in space, should have the transformation properties of a covariant pseudo vector if we assume its skew tensor representation (2.34) to be an ordinary tensor density.

A similar statement can be made about the magnetic induction in free space. Consider the magnetic induction as a skew symmetric covariant tensor†

$$B_{\lambda'\nu'} = A^{\lambda\ \nu}_{\lambda'\nu'} B_{\lambda\nu}\,. \tag{2.36}$$

† See § 3.4 and § 7.1 for the transformations of the magnetic field and the magnetic induction as given by (2.34) and (2.36).

A contraction over λ and ν with (2.32) then yields the dual representation as a pseudo-vector density of weight + 1

$$\tilde{\mathfrak{B}}^{\sigma} = \tfrac{1}{2}\tilde{\mathfrak{E}}^{\lambda\nu\sigma}B_{\lambda\nu}\,. \tag{2.37}$$

The permeability tensor connecting (2.37) and (2.35), however, is an ordinary density of weight + 1

$$\tilde{\mathfrak{B}}^{\lambda} = \mu^{\lambda\nu}\tilde{H}_{\nu}\,. \tag{2.38}$$

The dual of $\mu^{\nu\lambda}$ is still an ordinary tensor, because it requires the application of (2.33) twice

$$\mu_{\lambda\nu\sigma} = \mu^{\rho\tau}\tilde{\mathrm{e}}_{\rho\lambda\nu}\tilde{\mathrm{e}}_{\tau\sigma\kappa}\,, \tag{2.39}$$

its weight, however, is − 1.

The tensor (2.39) of valence four relates the magnetic field and magnetic induction in their skew tensor representation

$$B_{\lambda\nu} = \tfrac{1}{2}\mu_{\lambda\nu\sigma\kappa}\mathfrak{H}^{\sigma\kappa}\,. \tag{2.40}$$

It is still possible, though, to obtain a pseudo-tensor representation of the permeability, by using a hybrid combination of the fields, e.g.,

$$B_{\lambda\nu} = \tilde{\mu}_{\lambda\nu\cdot}^{\cdot\cdot\sigma}\tilde{H}_{\sigma}\,. \tag{2.41}$$

It is clear from the formulae (2.31) . . . (2.41) that tensors can be equivalently represented by pseudo tensors and conversely. There is nothing very profound about this. The fundamental physical issue, however, is whether the postulated combination of transformation behavior, as given by (2.34) and (2.36), is an admissible physical choice. We may illustrate this in the following diagram:

	$B_{\sigma\kappa}$	$\tilde{B}_{\sigma\kappa}$	
$\mathfrak{H}^{\lambda\nu}$	admissible choice	not admissible choice	$\tilde{H}_\sigma$
$\tilde{\mathfrak{H}}^{\lambda\nu}$	not admissible choice	admissible choice	H_σ
	$\tilde{\mathfrak{B}}^\nu$	$\mathfrak{B}^\nu$	

(2.42)

The diagram shows that there are four admissible and equivalent representations for the permeability tensor, i.e.,

$$\begin{array}{ll} \mu_{\lambda\nu\sigma\kappa} & \text{weight } -1 \\ \mu^{\nu\sigma} & \text{weight } +1 \\ \left.\begin{array}{l} \tilde{\mu}_{\lambda\nu\cdot}^{\cdot\cdot\sigma} \\ \tilde{\mu}_{\cdot\sigma\kappa}^{\nu\cdot\cdot} \end{array}\right\} & \text{weight zero .} \end{array} \tag{2.43}$$

One obtains the representation corresponding to the off-diagonal squares if one interchanges the tensors and pseudo tensors

$$\begin{array}{ll} \tilde{\mu}_{\lambda\nu\sigma\kappa} & \text{weight } -1 \\ \tilde{\mu}^{\nu\sigma} & \text{weight } +1 \\ \left.\begin{array}{l} \mu_{\lambda\nu\cdot}^{\cdot\cdot\sigma} \\ \mu_{\cdot\sigma\kappa}^{\nu\cdot\cdot} \end{array}\right\} & \text{weight zero .} \end{array} \tag{2.44}$$

The representation (2.44) is not admissible, because the

permeability would vanish for a coordinate inversion†. Media admitting a coordinate inversion have a center of symmetry. The representation (2.44) would, therefore, lead to the contradictory conclusion that there would be no magnetic permeability in a medium with a center of symmetry.

The admissible representations (2.43) of the permeability still leaves a choice between the two diagonal squares of diagram (2.42), corresponding to two different representations of the magnetic field and the magnetic induction. It is common practice to use the field representation corresponding to the diagonal square in the upper left corner of (2.42), i.e.,

$$\boxed{\mathfrak{H}^{\lambda\nu} \quad \text{or} \quad \tilde{H}_\sigma \quad \text{with} \quad B_{\sigma\kappa} \quad \text{or} \quad \tilde{\mathfrak{B}}^\nu} \tag{2.45}$$

The other representation of the magnetic field, which is admissible but not commonly used, corresponds to the lower right hand corner of (2.42). It is

$$\tilde{\mathfrak{H}}^{\lambda\nu} \quad \text{or} \quad H_\sigma \quad \text{with} \quad \tilde{B}_{\sigma\kappa} \quad \text{or} \quad \mathfrak{B}^\nu . \tag{2.46}$$

The tensors and pseudo tensors in (2.45) and (2.46) alternate. This affects what is usually called the sense of orientation associated with the geometrical images of the fields††. An inspection of these geometrical images then strongly suggests the use of (2.45) as the physically most direct and realistic choice.

A diagram similar to (2.42) can be made for the four dimensional case with the fields $\mathfrak{G}^{\lambda\nu}$ and $F_{\sigma\kappa}$. The pseudo per-

† It is a consequence of the Neumann principle of crystal physics that a medium with a symmetry center allows a coordinate inversion $x^{\lambda'} = -x^\lambda$, $(\lambda' = \lambda)$.

†† SCHOUTEN [1951] Chapter II, § 8, uses the term *W* tensor for pseudo tensor, after H. WEYL [1921] who introduced densities.

mutation fields have now four indices. They can be used to produce the dual representations of the electromagnetic field tensors. These "duals", as to be expected, are still tensors of the skew symmetric type

$$\tilde{\mathfrak{F}}^{\lambda\nu} = \tilde{\mathfrak{E}}^{\lambda\nu\sigma\kappa}F_{\sigma\kappa} \qquad \lambda, \nu \ldots = 0, 1, 2, 3 \tag{2.47}$$

$$\tilde{G}_{\lambda\nu} = \tilde{\mathfrak{e}}_{\lambda\nu\sigma\kappa}\mathfrak{G}^{\sigma\kappa}. \tag{2.48}$$

The diagram corresponding to (2.42) becomes

	$F_{\sigma\kappa}$	$\tilde{F}_{\sigma\kappa}$	
$\mathfrak{G}^{\lambda\nu}$	admissible choice	not admissible choice	$\tilde{G}_{\lambda\nu}$
$\tilde{\mathfrak{G}}^{\lambda\nu}$	not admissible choice	admissible choice	$G_{\lambda\nu}$
	$\tilde{\mathfrak{F}}^{\sigma\kappa}$	$\mathfrak{F}^{\sigma\kappa}$	

(2.49)

The reader will not have any difficulty confirming that the admissible choice for (2.49), compatible with the three-dimensional representation (2.45), is given by

$$\boxed{\mathfrak{G}^{\lambda\nu} \text{ or } \tilde{G}_{\lambda\nu} \text{ with } F_{\sigma\kappa} \text{ or } \tilde{\mathfrak{F}}^{\sigma\kappa}} \tag{2.50}$$

The representation (2.50) yields a connecting quantity [see (2.12)] for $\mathfrak{G}$ and F

$$\chi^{\lambda\nu\sigma\kappa}. \tag{2.51}$$

The quantity (2.51) is an ordinary tensor density, its "dielectric" and "magnetic" parts [see (2.14)] can never vanish as the result of a time or space inversion. The off-diagonal squares in (2.14), however, do vanish as the result of a space or time inversion. This conclusion should be modified if one admits a tensor (2.51) with complex elements. Details of the physical phenomena associated with a complex tensor (2.51) are discussed in chapters 6 and 8. It is then convenient to remark that the off-diagonal matrix $\tilde{\gamma}$ in (2.14) can be interpreted as a three-dimensional pseudo tensor of valence two

$$\tfrac{1}{2}\chi^{o\nu\sigma\kappa}\ \tilde{e}_{\sigma\kappa\tau} = \tilde{\gamma}_{\nu\tau} \qquad \nu,\sigma,\kappa,\tau = 1,2,3, \tag{2.52}$$

where the superscript zero in (2.52) is a time label.

CHAPTER 3

NATURAL INVARIANCE OF THE MAXWELL EQUATIONS

1. DIFFERENT FORMS OF THE MAXWELL EQUATIONS

The earlier literature usually presents the Maxwell equations in the Gaussian system of mixed units. For free space they are given by the following set of equations

$$\begin{aligned} &\underset{\text{e.s.u.}}{\operatorname{curl} \boldsymbol{E}} = \underset{\text{e.m.u.}}{-\frac{1}{c}\frac{\partial}{\partial t}\boldsymbol{H}}\,, \qquad &&\underset{\text{e.m.u.}}{\operatorname{div} \boldsymbol{H} = 0}\,, \\ &\underset{\text{e.m.u.}}{\operatorname{curl} \boldsymbol{H}} = \frac{1}{c}\underset{\text{e.s.u.}}{\left(\frac{\partial}{\partial t}\boldsymbol{E} + 4\pi \boldsymbol{s}\right)}\,, \qquad &&\underset{\text{e.s.u.}}{\operatorname{div} \boldsymbol{E}} = \underset{\text{e.s.u.}}{4\pi\rho}\,. \end{aligned} \tag{3.1}$$

The choice of units has been indicated underneath the relevant members of the equations (e.m.u. = electromagnetic units, e.s.u. = electrostatic units), c is the empty space light velocity. It is customary to use this same set of equations for microphysical applications. The current density $\boldsymbol{s}$ is then given by the charge densities times the drift velocities of the charges.

It should be kept in mind that the time derivatives in (3.1) are to be regarded as partial derivatives (see § 3.4).

Assuming matter-free space (i.e., $\boldsymbol{s} = 0$, $\rho = 0$) and eliminating one of the field vectors $\boldsymbol{E}$ or $\boldsymbol{H}$ one finds that each field vector satisfies the d'Alembertian wave equations

$$\left(\frac{\partial}{\partial x^2} + \frac{\partial}{\partial y^2} + \frac{\partial}{\partial z^2}\right)\begin{Bmatrix}\boldsymbol{E}\\ \boldsymbol{H}\end{Bmatrix} = \frac{1}{c^2}\frac{\partial^2}{\partial t^2}\begin{Bmatrix}\boldsymbol{E}\\ \boldsymbol{H}\end{Bmatrix}. \tag{3.2}$$

The wave equation (3.2) is in a form that makes the Lorentz invariance quite conspicuous.

The equations (3.1) can be regarded as a decomposition of the second order wave equation (3.2) into four first order equations. The above type of decomposition of the wave equation, as a matter of fact, is not unique. The equations (3.1) are commonly called the field equations in contradistinction with the wave equation (3.2).

A further distinction between field and wave equations in electromagnetics is brought to bear if one considers a material medium with magnetic permeability and electric permittivity. Experimental observation then suggests the introduction of two new field vectors known as dielectric displacement $\boldsymbol{D}$ and magnetic induction $\boldsymbol{B}$. The field equations for a material medium are known to be

$$\begin{aligned} &\underset{\text{e.s.u.}}{\operatorname{curl} \boldsymbol{E}} = -\underset{\text{e.m.u.}}{\frac{1}{c}\frac{\partial}{\partial t}\boldsymbol{B}}\,, &\qquad &\underset{\text{e.m.u.}}{\operatorname{div} \boldsymbol{B}} = 0\,, \\ &\underset{\text{e.m.u.}}{\operatorname{curl} \boldsymbol{H}} = \frac{1}{c}\left(\underset{\text{e.s.u.}}{\frac{\partial}{\partial t}\boldsymbol{D} + 4\pi \boldsymbol{s}}\right), &\qquad &\underset{\text{e.s.u.}}{\operatorname{div} \boldsymbol{D}} = 4\pi\rho\,. \end{aligned} \tag{3.3}$$

Lorentz showed that the equations (3.3) can be derived from the set (3.1) if one makes a number of very reasonable and acceptable assumptions about the electric structure of matter (ROSENFELD [1951]).

For a derivation of a wave equation one needs now additional information in the form of the so-called constitutive equations. Their simplest form for a linear, isotropic medium is

$$\boldsymbol{D} = \underset{r}{\varepsilon}\,\boldsymbol{E}, \qquad \boldsymbol{B} = \underset{r}{\mu}\,\boldsymbol{H}, \tag{3.4}$$

with $\underset{s}{\varepsilon}$ and $\underset{r}{\mu}$ as two dimensionless constants known as relative permittivity and relative permeability.

The equations (3.3) and (3.4) lead to a wave equation of the form

$$\left(\frac{\partial}{\partial x^2}+\frac{\partial}{\partial y^2}+\frac{\partial}{\partial z^2}\right)=\frac{\varepsilon_r \mu_r}{c^2}\frac{\partial^2}{\partial t^2}. \tag{3.5}$$

The equation (3.5) unlike (3.2) is not invariant under Lorentz transformations. In point of fact one would not expect it to be invariant under Lorentz transformations, because wave phenomena in moving media are known not to be independent of the motion of the medium (e.g., the Fresnel-Fizeau effect).

The insertion of matter into the electromagnetic field, therefore, requires us to reconsider our notions about Lorentz invariance. The reorientation of concepts is really the transition from invariance to covariance†, Lorentz "invariance" in the case of equation (3.2) informs us about the fact that wave propagation in matter-free space is independent of the motion of the observer. Lorentz "covariance" in the case of (3.5) should inform us how (3.5) should be modified to account for the phenomena in a moving medium, in conjunction with (3.3) and (3.4) (chapter 8).

The simplest and most popular explanation of the Fresnel-Fizeau effect is given by means of the well-known relativistic addition theorem of velocities. This theorem holds for group velocities only, hence applications to anisotropic and dispersive media, where group- and phase velocity are not identical, should be considered with due discrimination. For a good understanding of covariance and electromagnetism, however, it is useful to pursue the details of covariance in (3.3) and (3.4). Once established they determine the behavior of (3.5)

† The words "invariance" and "covariance" are very loosely manipulated in the physical literature. We will attempt to be consistent about their usage, though we do not guarantee that there will not be a slip somewhere in the text.

for a medium in motion. The question thus arises: are (3.3) and (3.4) both affected by a Lorentz transformation, or one of them only? The answer is immediately suggested by the fact that (3.3) should be valid for any medium, including matter-free space, while (3.4) characterizes the medium. Hence the "field equations" (3.3) should be form invariant under a Lorentz transformation and the constitutive equations should be subject to certain specific changes to account for the properties and the state of motion of the medium.

In other words, the introduction of the two new field variables $\boldsymbol{D}$ and $\boldsymbol{B}$ enables one to delineate further the characteristics of the field equations (3.3) and the constitutive equations (3.4). The Maxwell-Lorentz equations (3.1), from the present viewpoint are a sort of special combination of field equations and constitutive equations which apply only to free space. However, it should not be forgotten that the equations (3.1) are instrumental in obtaining a microphysical foundation for the more general field equations (3.3) by means of the electron theory (ROSENFELD [1951]).

A *formal* distinction between $\boldsymbol{E}$, $\boldsymbol{D}$ and $\boldsymbol{H}$, $\boldsymbol{B}$ in empty space becomes desirable if one considers the use of coherent units in the Maxwell equations instead of the mixed electrostatic and electromagnetic units used in (3.1) and (3.3).

For a coherent system of units the Maxwell equations can be brought into a form which is free of any constants or parameters which are reminiscent of the properties of any one particular medium. The equations (3.1) and (3.3) then become

$$\begin{aligned} \operatorname{curl} \boldsymbol{E} &= -\frac{\partial}{\partial t} \boldsymbol{B}, \qquad & \operatorname{div} \boldsymbol{B} &= 0, \\ \operatorname{curl} \boldsymbol{H} &= \frac{\partial}{\partial t} \boldsymbol{D} + \boldsymbol{s}, \qquad & \operatorname{div} \boldsymbol{D} &= \rho. \end{aligned} \tag{3.6}$$

The corresponding constitutive equations are: for empty space

$$\boldsymbol{D} = \underset{0}{\varepsilon}\boldsymbol{E}\,, \qquad \boldsymbol{B} = \underset{0}{\mu}\boldsymbol{H}\,, \tag{3.7}$$

and for a linear, and stationary, isotropic, material medium

$$\boldsymbol{D} = \underset{r}{\varepsilon}\underset{0}{\varepsilon}\boldsymbol{E}\,, \qquad \boldsymbol{B} = \underset{r}{\mu}\underset{0}{\mu}\boldsymbol{H}\,. \tag{3.8}$$

The values of the constants $\underset{0}{\varepsilon}$ and $\underset{0}{\mu}$ are for the three most important coherent systems

$$\begin{aligned} &\underset{0}{\varepsilon} = \frac{1}{4\pi}\,, && \underset{0}{\mu} = \frac{4\pi}{c^2}\,, && c \text{ in cm/sec}\,, && \text{e.s.u.} \\ &\underset{0}{\varepsilon} = \frac{1}{4\pi c^2}\,, && \underset{0}{\mu} = 4\pi\,, && c \text{ in cm/sec}\,, && \text{e.m.u.} \\ &\underset{0}{\varepsilon} = \frac{10^7}{4\pi c^2}\,, && \underset{0}{\mu} = \frac{4\pi}{10^7}\,, && c \text{ in m/sec}\,, && \text{m.k.s.} \end{aligned} \tag{3.9}$$

The m.k.s. system was proposed early in the century by G. Giorgi to maintain the practical units of potential and current (volt and ampere). It was accepted by the International Electrotechnical Committee in 1935. The present situation is that the Gaussian mixed system of units is still most popular with the physical theorists, while the m.k.s. system is now almost in universal use for macroscopic technical application of Maxwell's theory.

We may not refrain from noting here that the m.k.s. system stresses the properties of the vacuum; it is therefore more suggestive for covariant generalizations of the constitutive equations. The point is that there is no real physical issue as yet, which explains the reluctance of the physicist to transcribe all the classical information into m.k.s. units. The $\underset{r}{\varepsilon}$ and the $\underset{r}{\mu}$ are the physically important parameters associated with the structure of matter. The $\underset{0}{\varepsilon}$ and $\underset{0}{\mu}$ are the parameters associated with the structure of the space-time manifold; for practical purposes they play physically a secondary role compared to the influence of the structure of matter.

The distinction between ***E***, ***D*** and ***H***, ***B*** will be a physical issue if general covariance becomes a major physical issue (chapter 9). In the present text we will maintain the distinction because, even if there is no purely physical justification then there still is a mathematical physical justification: e.g., a coherent and unambiguous treatment of nonuniform media and curvilinear coordinates (chapter 7).

An investigation of the invariance of the equations (3.1) led to the now famous group of Lorentz transformations. The Lorentz transformations were the first transformations in physics where time joined the ranks of the position coordinates in the sense that all of them can be subject to a linear transformation. If time is subject to a transformation for uniform translations, as is the case for Lorentz transformations, then it is quite certain that time should be subject to a transformation for nonuniform translations. Whether this transformation will deviate considerably from what may already be expected on ground of a Lorentz transformation using the instantaneous velocity is at present not known†. So far the Lorentz transformations are the most important space-time transformations which have a well specified meaning in physics. There is no doubt that there are other space-time transformations corresponding to more complex motions of the frames of reference. These, however, have not been specified in any complete sense so that they lead to a set of well integrated and accepted conclusions in physics.

Our lack of knowledge of nonuniform space-time transformations is no reason to ignore the necessity of general

† Speculations about this subject lead to some notorious time paradoxes of relativity.

covariance. Any formulation short of general covariance will almost certainly not provide an answer to physical problems associated with nonuniform space-time systems. The important point is whether one can find a meaningful and unambiguous generally covariant formulation, without having a detailed physical knowledge of nonuniform space-time transformations. Physically significant checks of generally covariant relations can always be made by using subgroups of P_4 which preserve absolute time. An example is discussed in § 3.4.

After the realization of a certain similarity between space coordinates and time, Minkowski was one of the first to investigate whether physical fields could be defined in a way so that they appear as fields in the four-dimensional space-time manifold. He discovered that electric and magnetic vectors can be pairwise combined into four-dimensional skew symmetric tensors according to the scheme:

$$\left.\begin{matrix}\boldsymbol{E}\\ \\ \boldsymbol{B}\end{matrix}\right\} \rightarrow F_{\lambda\nu} = -F_{\nu\lambda} \quad \text{and} \quad \left.\begin{matrix}\boldsymbol{D}\\ \\ \boldsymbol{H}\end{matrix}\right\} \rightarrow \mathfrak{G}^{\lambda\nu} = -\mathfrak{G}^{\nu\lambda}, \tag{3.10}$$

$$\left.\begin{matrix} \\ \\ \boldsymbol{s}\end{matrix}\right\} \rightarrow \mathfrak{c}^{\lambda}, \qquad \lambda, \nu = 0, 1, 2, 3 .$$

The zero denotes the time label. The component-wise $(2+2\times 3)$ Maxwell equations then are represented by the (2×4) Minkowski equations

$$\partial_{[\kappa}F_{\lambda\nu]} = 0 , \tag{3.11}$$

$$\partial_{\nu}\mathfrak{G}^{\lambda\nu} = \mathfrak{c}^{\lambda}, \qquad \kappa, \lambda, \nu = 0, 1, 2, 3 . \tag{3.12}$$

The equation (3.11) becomes, if the alternation is expanded as specified in (1.17),

		Constitutive equations	
	Maxwell equations	Free space	Isotrop nonconduc matte
A (3.1)	$\text{curl } \boldsymbol{E} = -\dfrac{1}{c}\dfrac{\partial}{\partial t}\boldsymbol{H}$ $\text{div } \boldsymbol{H} = 0$ $\text{curl } \boldsymbol{H} = \dfrac{1}{c}\left(\dfrac{\partial}{\partial t}\boldsymbol{E} + 4\pi \boldsymbol{s}\right)$ $\text{div } \boldsymbol{E} = 4\pi\rho$	None	None
B (3.3)	$\text{curl } \boldsymbol{E} = -\dfrac{1}{c}\dfrac{\partial}{\partial t}\boldsymbol{B}$ $\text{div } \boldsymbol{B} = 0$ $\text{curl } \boldsymbol{H} = \dfrac{1}{c}\left(\dfrac{\partial}{\partial t}\boldsymbol{D} + 4\pi \boldsymbol{s}\right)$ $\text{div } \boldsymbol{D} = 4\pi\rho$	$\boldsymbol{D} = \boldsymbol{E}$ $\boldsymbol{B} = \boldsymbol{H}$	$\boldsymbol{D} = \varepsilon_r$ $\boldsymbol{B} = \mu_r$
C (3.6)	$\text{curl } \boldsymbol{E} = -\dfrac{\partial}{\partial t}\boldsymbol{B}$ $\text{div } \boldsymbol{B} = 0$ $\text{curl } \boldsymbol{H} = \dfrac{\partial}{\partial t}\boldsymbol{D} + \boldsymbol{s}$ $\text{div } \boldsymbol{D} = \rho$	$\boldsymbol{D} = \varepsilon_0 \boldsymbol{E}$ $\boldsymbol{B} = \mu_0 \boldsymbol{H}$	$\boldsymbol{D} = \varepsilon_r\varepsilon_0$ $\boldsymbol{B} = \mu_r\mu_0$

LE 2

Identifications in four-space

Fields	Coordinates
$\begin{pmatrix} 0 & F_{01} & F_{02} & F_{03} \\ F_{10} & 0 & F_{12} & F_{13} \\ F_{20} & F_{21} & 0 & F_{23} \\ F_{30} & F_{31} & F_{32} & 0 \end{pmatrix}$, $\begin{pmatrix} 0 & \mathfrak{G}^{01} & \mathfrak{G}^{02} & \mathfrak{G}^{03} \\ \mathfrak{G}^{10} & 0 & \mathfrak{G}^{12} & \mathfrak{G}^{13} \\ \mathfrak{G}^{20} & \mathfrak{G}^{21} & 0 & \mathfrak{G}^{23} \\ \mathfrak{G}^{30} & \mathfrak{G}^{31} & \mathfrak{G}^{32} & 0 \end{pmatrix} = \begin{pmatrix} 0 & iE_1 & iE_2 & iE_3 \\ -iE_1 & 0 & H_3 & -H_2 \\ -iE_2 & -H_3 & 0 & H_1 \\ -iE_3 & H_2 & -H_1 & 0 \end{pmatrix}$; $\mathfrak{c}^0 = -i\rho$, $\mathfrak{c}^1 = s_1$, $\mathfrak{c}^2 = s_2$, $\mathfrak{c}^3 = s_3$	$x_0 = -ict$, $x_1 = x$, $x_2 = y$, $x_3 = z$
$\begin{pmatrix} 0 & F_{01} & F_{02} & F_{03} \\ F_{10} & 0 & F_{12} & F_{13} \\ F_{20} & F_{21} & 0 & F_{23} \\ F_{30} & F_{31} & F_{32} & 0 \end{pmatrix} = \begin{pmatrix} 0 & -E_1 & -E_2 & -E_3 \\ E_1 & 0 & \frac{1}{c}B_3 & -\frac{1}{c}B_2 \\ E_2 & -\frac{1}{c}B_3 & 0 & \frac{1}{c}B_1 \\ E_3 & \frac{1}{c}B_2 & -\frac{1}{c}B_1 & 0 \end{pmatrix}$; $\begin{pmatrix} 0 & \mathfrak{G}^{01} & \mathfrak{G}^{02} & \mathfrak{G}^{03} \\ \mathfrak{G}^{10} & 0 & \mathfrak{G}^{12} & \mathfrak{G}^{13} \\ \mathfrak{G}^{20} & \mathfrak{G}^{21} & 0 & \mathfrak{G}^{23} \\ \mathfrak{G}^{30} & \mathfrak{G}^{31} & \mathfrak{G}^{32} & 0 \end{pmatrix} = \begin{pmatrix} 0 & \frac{1}{c}D_1 & \frac{1}{c}D_2 & \frac{1}{c}D_3 \\ -\frac{1}{c}D_1 & 0 & H_3 & -H_2 \\ -\frac{1}{c}D_2 & -H_3 & 0 & H_1 \\ -\frac{1}{c}D_3 & H_2 & -H_1 & 0 \end{pmatrix}$; $\mathfrak{c}^0 = \rho$, $\mathfrak{c}^1 = s_1$, $\mathfrak{c}^2 = s_2$, $\mathfrak{c}^3 = s_3$	$x_0 = t$, $x_1 = x$, $x_2 = y$, $x_3 = z$
$\begin{pmatrix} 0 & F_{01} & F_{02} & F_{03} \\ F_{10} & 0 & F_{12} & F_{13} \\ F_{20} & F_{21} & 0 & F_{23} \\ F_{30} & F_{31} & F_{32} & 0 \end{pmatrix} = \begin{pmatrix} 0 & -E_1 & -E_2 & -E_3 \\ E_1 & 0 & B_3 & -B_2 \\ E_2 & -B_3 & 0 & B_1 \\ E_3 & B_2 & -B_1 & 0 \end{pmatrix}$; $\begin{pmatrix} 0 & \mathfrak{G}^{01} & \mathfrak{G}^{02} & \mathfrak{G}^{03} \\ \mathfrak{G}^{10} & 0 & \mathfrak{G}^{12} & \mathfrak{G}^{13} \\ \mathfrak{G}^{20} & \mathfrak{G}^{21} & 0 & \mathfrak{G}^{23} \\ \mathfrak{G}^{30} & \mathfrak{G}^{31} & \mathfrak{G}^{32} & 0 \end{pmatrix} = \begin{pmatrix} 0 & D_1 & D_2 & D_3 \\ -D_1 & 0 & H_3 & -H_2 \\ -D_2 & -H_3 & 0 & H_1 \\ -D_3 & H_2 & -H_1 & 0 \end{pmatrix}$; $\mathfrak{c}^0 = \rho$, $\mathfrak{c}^1 = s_1$, $\mathfrak{c}^2 = s_2$, $\mathfrak{c}^3 = s_3$	$x_0 = t$, $x_1 = x$, $x_2 = y$, $x_3 = z$

$$\frac{1}{3!}\left(\partial_\kappa F_{\lambda\nu} + \partial_\nu F_{\kappa\lambda} + \partial_\lambda F_{\nu\kappa} - \partial_\kappa F_{\nu\lambda} - \partial_\nu F_{\lambda\kappa} - \partial_\lambda F_{\kappa\nu}\right)$$

$$= \tfrac{1}{3}\left(\partial_\kappa F_{\lambda\nu} + \partial_\nu F_{\kappa\lambda} + \partial_\lambda F_{\nu\kappa}\right) = 0\,.$$

For nonrationalized units equation (3.12) should be written with the factor 4π,

$$\partial_\nu \mathfrak{G}^{\lambda\nu} = 4\pi \mathfrak{c}^\lambda\,. \tag{3.12a}$$

The equations (3.11) and (3.12), as will be shown in the next section, are form invariant for general space-time transformations, if one adopts the transformation behavior of the field as quoted in table 1 of chapter 2. The early Minkowski renditions of (3.11) and (3.12) did not obey quite such general invariance properties for a number of reasons depending on the details of the identification (3.10). There is a mild chaos in the literature regarding to what one likes to consider as the basic form of the Minkowski equations and to what should be considered as the best identification in four-space in the sense of (3.10).

A choice of the most important identification possibilities, in the sense of (3.10), is summarized in table 2. It corresponds to the forms (3.1), (3.3) and (3.6) of the Maxwell equations: rows A, B and C. A single identification for F and $\mathfrak{G}$ suffices in the case of row A, provided one makes the notorious time coordinate identification $x_0 = ict$. The free-space constitutive equations are then implicit in the "field equations" and the covariance is restricted to the Lorentz group or rather the four-dimensional orthogonal group with the "adulterated" metric $x_0^2 + x_1^2 + x_2^2 + x_3^2$.

The identifications given in rows B and C lead to the generally covariant field equations (3.11) and (3.12) [(3.12a) for B and (3.12) for C]. The free-space constitutive equations, however, are restricted to the Lorentz group.

In the present text we shall adopt the identification corresponding to row C of table 2.

2. NATURAL INVARIANCE OF THE MAXWELL EQUATIONS

The Minkowski equations (3.11) and (3.12) of the previous section retain their form for arbitrary continuous space-time transformations, if one agrees on the transformation behavior of $F_{\lambda\nu}$ and $\mathfrak{G}^{\lambda\nu}$ as given in table 1 of chapter 2. The proof is straightforward and simple. The natural invariance of (3.11) implies that the form $\partial_{[\kappa}F_{\lambda\nu]}$ transforms as a covariant trivector (alternating tensor, valence 3)

$$\partial_{[\kappa'}F_{\lambda'\nu']} = A^{\kappa\,\lambda\,\nu}_{\kappa'\lambda'\nu'}\partial_{[\kappa}F_{\lambda\nu]}\,. \tag{3.13}$$

Hence if $\partial_{[\kappa}F_{\lambda\nu]} = 0$, then $\partial_{[\kappa'}F_{\lambda'\nu']} = 0$.

The fields $F_{\lambda'\nu'}$ and $F_{\lambda\nu}$ are related by

$$F_{\lambda'\nu'} = A^{\lambda\,\nu}_{\lambda'\nu'}F_{\lambda\nu}\,. \tag{3.14}$$

Differentiation of (3.14) with respect to $x^{\kappa'}$ and alternating over κ', λ' and ν' yields

$$\partial_{[\kappa'}F_{\lambda'\nu']} = \partial_{[\kappa'}(A^{\lambda\,\nu}_{\lambda'\nu']}F_{\lambda\nu})\,. \tag{3.15}$$

Expanding the differentiation over the individual coefficients of the right-hand member of (3.15) gives

$$\partial_{[\kappa'}F_{\lambda'\nu']} = A^{\lambda\,\nu}_{[\lambda'\nu'}\partial_{\kappa']}F_{\lambda\nu} + F_{\lambda\nu}A^{\nu}_{[\nu'}\partial_{\kappa'}A^{\lambda}_{\lambda']} + F_{\lambda\nu}A^{\lambda}_{[\lambda'}\partial_{\kappa'}A^{\nu}_{\nu']}\,. \tag{3.16}$$

The second and third terms in the right-hand member of (3.16) vanish for holonomic frames of reference, because the expressions

$$\partial_{\kappa'}A^{\lambda}_{\lambda'} \quad \text{and} \quad \partial_{\kappa'}A^{\nu}_{\nu'}$$

alternated over κ', λ' and κ', ν' vanish due to (1.13); thus (3.16) is reduced to

$$\partial_{[\kappa'}F_{\lambda'\nu']} = A^{\lambda\,\nu}_{[\lambda'\nu'}\partial_{\kappa']}F_{\lambda\nu}\,. \tag{3.17}$$

Furthermore, keeping in mind that

$$\frac{\partial}{\partial x^{\kappa'}} = \frac{\partial x^{\kappa}}{\partial x^{\kappa'}} \frac{\partial}{\partial x^{\kappa}},$$

or in the present notation

$$\partial_{\kappa'} = A^{\kappa}_{\kappa'} \partial_{\kappa}, \tag{3.18}$$

equation (3.17) can be written as

$$\partial_{[\kappa'} F_{\lambda'\nu']} = A^{\lambda\,\nu\,\kappa}_{[\lambda'\nu'\kappa']} \partial_{\kappa} F_{\lambda\nu}. \tag{3.19}$$

Because of the summation over $\lambda\nu\kappa$ the alternation sign [ ] can be applied either to $\lambda'\nu'\kappa'$ or $\lambda\nu\kappa$. Therefore

$$\partial_{[\kappa'} F_{\lambda'\nu']} = A^{\kappa\,\lambda\,\nu}_{\kappa'\lambda'\nu'} \partial_{[\kappa} F_{\lambda\nu]}{}^{\dagger}$$

which is identical with (3.13) q.e.d.

We may expect the proof of the natural invariance of the second Minkowski equations (3.12) to be associated with the previous proof on basis of the duality provided by the permutation fields defined by (1.20) and (1.21). A covariant bivector in a four-dimensional manifold always has associated with it a contravariant bivector density by means of the relation

$$\mathfrak{G}^{\kappa\sigma} = \mathfrak{E}^{\kappa\sigma\lambda\nu} F_{\lambda\nu}. \tag{3.20}$$

Differentiation with respect to x^{σ} and contraction over σ gives

$$\partial_{\sigma} \mathfrak{G}^{\kappa\sigma} = \mathfrak{E}^{\kappa\sigma\lambda\nu} \partial_{\sigma} F_{\lambda\nu}, \tag{3.21}$$

† Please note that the sequence of indices $\lambda\nu\kappa$ in $A^{\lambda\,\nu\,\kappa}_{\lambda'\nu'\kappa'}$ can be rearranged if simultaneously $\lambda'\nu'\kappa'$ are submitted to the same rearrangement.

remembering that the permutation fields have constant components equal to ± 1.

The summation over the alternating indices $\sigma\lambda\nu$ implies that only the terms $\partial_{[\sigma}F_{\lambda\nu]}$ contribute in the summation. Hence

$$\partial_\sigma \mathfrak{G}^{\kappa\sigma} = \mathfrak{E}^{\kappa\sigma\lambda\nu}\partial_{[\sigma}F_{\lambda\nu]}\,. \tag{3.22}$$

We just proved that $\partial_{[\sigma}F_{\lambda\nu]}$ has transformation properties of a tensor. The permutation field is a tensor density. Therefore, the left hand member of (3.22) should transform as a contravariant vector density of weight $+1$, i.e.

$$\partial_{\sigma'}\mathfrak{G}^{\kappa'\sigma'} = |\Delta|^{-1} A_\kappa^{\kappa'}\partial_\sigma \mathfrak{G}^{\kappa\sigma}\,. \tag{3.23}$$

It is immaterial for the sake of the mathematical argument that the equation (3.20) is an empty statement from a point of view of physics. The equation (3.20) has the appearance of a somewhat peculiar constitutive equation. It contains, however, no physical information about the medium. The physically significant constitutive relations between $\mathfrak{G}^{\lambda\nu}$ and $F_{\lambda\nu}$ are not of the type (3.20). They will be discussed in chapter 6.

For the sake of completeness and for a little practising experience in the kernel-index method, we may give an independent proof of the natural invariance of (3.12).

The transformation of the field $\mathfrak{G}^{\lambda\nu}$ according to table 1 of chapter 2 is given by

$$\mathfrak{G}^{\lambda'\nu'} = |\Delta|^{-1} A_{\lambda\,\nu}^{\lambda'\nu'}\mathfrak{G}^{\lambda\nu}. \tag{3.24}$$

Taking the derivative with respect to $x^{\nu'}$ (summation over ν') and expanding the differentiation over the factors of the right-hand member of (3.18) yields

$$\partial_{\nu'}\mathfrak{G}^{\lambda'\nu'} = |\Delta|^{-1} A_{\lambda\,\nu}^{\lambda'\nu'}\partial_{\nu'}\mathfrak{G}^{\lambda\nu}$$

$$+\,\mathfrak{G}^{\lambda\nu}\{|\Delta|^{-1}A_\lambda^{\lambda'}\partial_{\nu'}A_\nu^{\nu'} + |\Delta|^{-1}A_\nu^{\nu'}\partial_{\nu'}A_\lambda^{\lambda'} + A_{\lambda\,\nu}^{\lambda'\nu'}\partial_{\nu'}|\Delta|^{-1}\}. \tag{3.25}$$

The first term in the right-hand member can be converted by means of (3.18)

$$| \Delta |^{-1} A^{\lambda'\nu'}_{\lambda\ \nu} \partial_{\nu'} \mathfrak{G}^{\lambda\nu} = | \Delta |^{-1} A^{\lambda'}_{\lambda} \partial_{\nu} \mathfrak{G}^{\lambda\nu} , \tag{3.26}$$

which is exactly the form we are looking for in accordance with (3.23), hence we expect the second right-hand term in (3.25) to vanish.

The derivative of the Jacobian Δ can be converted according to

$$\partial_\nu \Delta^{-1} = - \Delta^{-1} \partial_\nu \ln \Delta = - \Delta^{-1} \frac{\partial \ln \Delta}{\partial A^{\kappa'}_{\kappa}} \partial_\nu A^{\kappa'}_{\kappa} ,$$

using (1.11)

$$\partial_\nu \Delta^{-1} = - \Delta^{-1} A^{\kappa}_{\kappa'} \partial_\nu A^{\kappa'}_{\kappa} . \tag{3.27}$$

Substitution of (3.26) and (3.27) into (3.25) yields

$$\partial_{\nu'} \mathfrak{G}^{\lambda'\nu'} = | \Delta |^{-1} A^{\lambda'}_{\lambda} \partial_\nu \mathfrak{G}^{\lambda\nu}$$

$$+ | \Delta |^{-1} \mathfrak{G}^{\lambda\nu} \{ A^{\lambda'\kappa}_{\lambda\ \nu'} \partial_\kappa A^{\nu'}_{\nu} + \partial_\nu A^{\lambda'}_{\lambda} - A^{\lambda'\kappa}_{\lambda\ \kappa'} \partial_\nu A^{\kappa'}_{\kappa} \} . \tag{3.28}$$

In (3.28) we have once more used (3.18) by writing

$$| \Delta |^{-1} A^{\lambda'\kappa}_{\lambda\ \nu'} \partial_\kappa A^{\nu'}_{\nu} \quad \text{for} \quad | \Delta |^{-1} A^{\lambda'}_{\lambda} \partial_{\nu'} A^{\nu'}_{\nu} .$$

The second line of terms in (3.28) vanishes for a holonomic system; namely the middle term drops out in the summation over $\lambda\nu$ because

$$\mathfrak{G}^{\lambda\nu} = - \mathfrak{G}^{\nu\lambda}$$

and

$$\partial_\nu A^{\lambda'}_{\lambda} = \partial_\lambda A^{\lambda'}_{\nu} . \tag{1.13}$$

The other two terms cancel

$$A^{\lambda'\kappa}_{\lambda\ \nu'} \partial_\kappa A^{\nu'}_{\nu} = A^{\lambda'\kappa}_{\lambda\ \kappa'} \partial_\nu A^{\kappa'}_{\kappa} ,^{\dagger}$$

† Note the replacement of the summation index ν' by κ'.

also because of the symmetry imposed by (1.11). Therefore, on holonomic frames of reference

$$\partial_{\nu'}\mathfrak{G}^{\lambda'\nu'} = |\Delta|^{-1}A_{\lambda}^{\lambda'}\partial_{\nu}\mathfrak{G}^{\lambda\nu}, \quad \text{q.e.d.} \tag{3.23}$$

The preceding tensor analytical digression completes the proof of natural invariance of the Maxwell equations for transitions between holonomic frames of reference. It should be noted that there is nothing really mathematically profound about the proofs. Only elementary notions of partial differentiation are required. There is a certain complexity which makes it important to tackle this sort of work with the right kind of notational machinery. It is believed that the kernel-index method asserts itself especially well in the treatment of transformation problems. Some individual exercise is indispensable, but this is true for a successful manipulation of any algorithm.

It is important to prove or check covariance properties of fundamental equations, but in addition it is essential to develop an eye for the recognition of covariance characteristics, as well as a feel for operations which do not disturb or upset covariance properties. Otherwise the manipulation of generally covariant expressions causes confusion and becomes a burden instead of being an asset in the delineation of physical and mathematical concepts. An appropriate and consistent notation is most helpful in attaining these goals.

The two natural invariant differential operations associated with the equations (3.11) and (3.12) are known as the generalized "curl" and "divergence" operation or "Stokian" derivatives†.

The "curl" is valid operating on any alternating covariant field; it increases the valence of the field with one alternating index.

† They are the derivatives occurring in the generalized integral laws of Stokes and Gauss, see SCHOUTEN [1951].

The "divergence" is valid operating on any alternating contravariant tensor density of weight $+1$; it decreases the valence of the field with one alternating index.

The natural invariance of the curl and divergence operations implies that they are independent of the structure of the manifold.

Suppose the manifold has a structure in the form of a Riemannian metric or a linear symmetric displacement. In either case one can use the covariant derivative associated with the metric or displacement. The covariant derivative provides a beloved recipe to transcribe physical equations into a covariant form. We may prove now that a transcription of the Minkowski equations (3.11) and (3.12) by means of the covariant derivative leads to the conclusion that the correction terms associated with the covariant derivative cancel. Hence covariant and ordinary derivative in the curl and divergence operations are identical.

For the equation (3.11) one obtains, replacing the ordinary derivative by the covariant derivative,

$$\nabla_{[\kappa} F_{\lambda\nu]} = \partial_{[\kappa} F_{\lambda\nu]} - 2F_{\sigma[\kappa} \Gamma^{\sigma}_{\lambda\nu]} \,. \tag{3.29}$$

The second term in the right-hand member cancels because of the alternation over the symmetric indices $\lambda\nu$ in $\Gamma^{\sigma}_{\lambda\nu}$

$$\Gamma^{\sigma}_{[\lambda\nu]} = 0 \quad [\text{see (1.15)}] \,. \tag{3.30}$$

Hence

$$\nabla_{[\kappa} F_{\lambda\nu]} = \partial_{[\kappa} F_{\lambda\nu]} \,. \tag{3.31}$$

A similar argument holds for equation (3.12)

$$\nabla_{\nu} \mathfrak{G}^{\lambda\nu} = \partial_{\nu} \mathfrak{G}^{\lambda\nu} + \mathfrak{G}^{\sigma\nu} \Gamma^{\lambda}_{\nu\sigma} + \mathfrak{G}^{\lambda\sigma} \Gamma^{\nu}_{\nu\sigma} - \Gamma^{\sigma}_{\sigma\nu} \mathfrak{G}^{\lambda\nu} \,. \tag{3.32}$$

The last term in the right-hand member stems from the density features of the field $\mathfrak{G}^{\lambda\nu}$. It cancels with the preceding term. The second term vanishes because of (3.30), and the summation over $\sigma\nu$ with the skew symmetric field $\mathfrak{G}^{\sigma\nu}$. Therefore

$$\nabla_\nu \mathfrak{G}^{\lambda\nu} = \partial_\nu \mathfrak{G}^{\lambda\nu}, \quad \text{q.e.d.} \tag{3.33}$$

It should be kept in mind that the relations (3.31) and (3.33) hold only for holonomic frames, because (3.30) is true on holonomic frames only. The absence of symmetry of the coefficients of linear displacement on anholonomic frames follows from the equations (2.27) and (2.28).

In the literature of differential geometry one occasionally encounters displacements with an inherent asymmetry not associated with the anholonomity of the reference systems. Their feasibility as physical entities will not be considered here. The point of primary importance was to stress the condition of holonomity for the natural invariance of the Maxwell equations. In contradistinction to analytical dynamics, anholonomic constraints have been hardly considered in field theory, although anholonomic systems of reference are quite common (see chapter 7).

The natural invariance of the Maxwell equations was first indicated by H. WEYL [1921]. Weyl emphasized explicitly the density aspects of the field in equation (3.12). One may have noticed that the natural invariance of (3.12) hinges on the fact that the field $\mathfrak{G}^{\lambda\nu}$ is a density of weight + 1. Other names associated with the natural invariance of equations (3.11) and (3.12) are E. Cartan and D. van Dantzig. A historical review and bibliography is given by E. WHITTAKER [1953].

3. THE POTENTIALS

The potentials can be regarded as fields which are particular solutions of the nonhomogeneous wave equation. The nonhomogeneous terms are the sources of the potential field. The most frequently occurring potentials and sources are

Source	Potential
Charge ρ	Scalar potential φ
Current $\boldsymbol{s}$	Vector potential $\boldsymbol{A}$
Electric dipole field $\boldsymbol{P}$	Hertz vector $\boldsymbol{\pi}$
Magnetic dipole field $\boldsymbol{M}$	Fitzgerald vector $\boldsymbol{F}$

If the wave equation is a d'Alembertian of the type (3.2) and if the source distribution is contained in a finite space volume τ, then the potentials are known to be related to the sources by expressions of the type

$$\text{potential} = \frac{1}{4\pi} \int_\tau \frac{\text{retarded source}}{r} \, d\tau \,, \tag{3.34}$$

r being the distance between source point and the position where the potential is observed. The time retardation of the source is given by r/c.

The three-dimensional scalar potential and vector potential are known to combine into a covariant four-vector A_λ with the identification

$$A_0 = -\varphi \quad \text{and} \quad (A_1, A_2, A_3) = \boldsymbol{A} \,. \tag{3.35}$$

The electric field and magnetic induction are related to the four-potential by the following naturally invariant equation

$$F_{\nu\lambda} = 2\partial_{[\nu} A_{\lambda]} \,. \tag{3.36}$$

One can prove in the same manner as in § 3.2 that its natural invariance holds for holonomic frames only.

Substitution of (3.36) in (3.11) shows that the latter equation is identically satisfied,

$$\partial_{[\kappa}\partial_{\nu}A_{\lambda]} = 0\,, \tag{3.37}$$

because of the alternation over the two differentiation indices κ, ν.

On similar grounds one shows that always a gradient vector can be added to A_λ without influencing the field $F_{\lambda\nu}$. One takes advantage of this arbitrariness in A_λ by submitting it to a "gauge" restriction known as the Lorentz condition†

$$\operatorname{div} \boldsymbol{A} + \underset{0}{\varepsilon}\underset{0}{\mu}\frac{\partial\varphi}{\partial t} = 0\,. \tag{3.38}$$

An inspection of (3.34) shows that the divergence relation (3.38) corresponds to the continuity relation of the sources

$$\operatorname{div} \boldsymbol{s} + \frac{\partial\rho}{\partial t} = 0\,. \tag{3.39}$$

The latter can be given a natural invariant four-dimensional form

$$\partial_\nu \mathfrak{c}^\nu = 0\,. \tag{3.39a}$$

The proof is similar to the proof of the natural invariance of equation (3.12). The invariance is again subject to the restriction of holonomic frames.

The Lorentz condition (3.38) can not be given a natural invariant form. This limitation is associated with the fact that the potentials (3.34) are solutions of wave equation (3.2) which itself is invariant under the Lorentz group only.

† The restriction implies that only gradients derived from scalars satisfying the homogeneous wave equation are admitted as additive gradient components of A_λ.

One encounters similar limitations for the Hertz and Fitzgerald vectors. The source fields for those are the electric polarization $\boldsymbol{P}$ and the magnetic polarization $\boldsymbol{M}$ per unit volume. These are related to the fields by the equations

$$\underset{0}{\varepsilon}\boldsymbol{E} = \boldsymbol{D} - \boldsymbol{P}$$
$$\frac{1}{\underset{0}{\mu}}\boldsymbol{B} = \boldsymbol{H} + \boldsymbol{M}. \tag{3.40}$$

From equation (3.40) one sees that if $\boldsymbol{D}$, $\boldsymbol{H}$ and $\boldsymbol{E}$, $\boldsymbol{B}$ combine into bivectors in four-space according to table 2, then $\boldsymbol{P}$ and $\boldsymbol{M}$ constitute a bivector in four-space with the identification

$$\mathfrak{M}^{01} = +P_1, \quad \mathfrak{M}^{02} = +P_2, \quad \mathfrak{M}^{03} = +P_3$$
$$\mathfrak{M}^{23} = -M_1, \quad \mathfrak{M}^{31} = -M_2, \quad \mathfrak{M}^{12} = -M_3. \tag{3.41}$$

For reasons of simplicity it is easier from here on to proceed on a general covariant basis. The free-space constants $\underset{0}{\varepsilon}$ and $\underset{0}{\mu}$ in (3.40) are restricted to Cartesian frames contained in the Lorentz group. The generally covariant rendition of (3.40) is

$$\mathfrak{G}^{\lambda\nu} - \mathfrak{M}^{\lambda\nu} = \tfrac{1}{2}\underset{0}{\chi}^{\lambda\nu\sigma\kappa}F_{\sigma\kappa}, \tag{3.40a}$$

where $\underset{0}{\chi}^{\lambda\nu\sigma\kappa}$ represents the free-space properties on general coordinates. The tensor density $\underset{0}{\chi}^{\lambda\nu\sigma\kappa}$ will be discussed in chapters 6 and following, we may mention here that it is skew symmetric in the indices $\lambda\nu$ and $\sigma\kappa$ and symmetric for a pair-wise interchange of same indices.

Suppose one takes the divergence of (3.40a) with respect to the index ν. Using (3.12) and (3.36) the resulting relation is

$$\mathfrak{c}^{\lambda} - \partial_{\nu}\mathfrak{M}^{\lambda\nu} = \partial_{\nu}\underset{0}{\chi}^{\lambda\nu\sigma\kappa}\partial_{\sigma}A_{\kappa} \tag{3.42}$$

↙	↘	↘
macroscopic charge and current density	microscopic charge and current density	the generalized d'Alembertian.

The equation (3.42) is the covariant analogue of a well known relation.

Returning to a Lorentz frame one can prove that the right-hand member of (3.42) reduces to the conventional d'Alembertian (3.2) by using the Lorentz condition (3.38)

$$\partial_{\nu}\underset{0}{\chi}^{\lambda\nu\sigma\kappa}\partial_{\sigma}A_{\kappa} = \frac{1}{\underset{0}{\mu}}\Box A^{\lambda}. \tag{3.43}$$

The A^{λ} are the contravariant components of the four potential, which are obtained by contraction with the metric

$$A^{\lambda} = g^{\lambda\nu}A_{\nu}, \tag{3.44}$$

with

$$g^{\lambda\nu} = \begin{pmatrix} \underset{0}{\varepsilon}\underset{0}{\mu} & & & \\ & -1 & & \\ & & -1 & \\ & & & -1 \end{pmatrix}.$$

Substituting (3.43) into (3.42) and assuming the macroscopic current and charge sources to be zero ($\mathfrak{c}^{\lambda} = 0$) gives the Lorentz invariant relation

$$-\partial_{\nu}\mathfrak{M}^{\lambda\nu} = \frac{1}{\underset{0}{\mu}}\Box A^{\lambda}. \tag{3.45}$$

The wave equation (3.45), with polarization sources only,

invites the substitution

$$A^{\lambda} = \partial_{\nu} Z^{\lambda\nu} \quad \text{with} \quad Z^{\lambda\nu} = - Z^{\nu\lambda}, \tag{3.46}$$

so that the potential $Z^{\lambda\nu}$ corresponds to the structure of the source $\mathfrak{M}^{\lambda\nu}$. A substitution of (3.46) in (3.45) and a removal of the divergence gives (except for a divergence-free part)

$$- \mathfrak{M}^{\lambda\nu} = \frac{1}{\mu_0} \Box Z^{\lambda\nu}. \tag{3.47}$$

The reader may confirm that the time components of $Z^{\lambda\nu}$ correspond to the Hertz vector $\boldsymbol{\pi}$, and the pure space components correspond to the Fitzgerald vector $\boldsymbol{F}$.

It is to be noted that the relations (3.43) . . . (3.47) are not natural invariants and neither is the Lorentz condition (3.42). They are, however, legitimate Lorentz invariants. The absence of natural invariance means that the relevant definitions and relations can not be formulated independent of the choice of coordinates and the structure of space-time.

The absence of natural invariance of course does not imply that these equations can not be written in a covariant form. Expression for curvilinear coordinates can be obtained by the customary procedures of covariant differentiation or by the method of anholonomic frames.

Details of the general covariance of the wave equation (3.42) will be discussed more elaborately in the chapters 6–9.

4. CORRELATION OF THREE- AND FOUR-DIMENSIONAL FORMS

The naturally invariant Minkowski equations (3.11) and (3.12) give rise to a naturally invariant three-dimensional form of the Maxwell equations:

$$2\partial_{[\nu}E_{\lambda]} = -\frac{\partial}{\partial t}B_{\nu\lambda}\,, \qquad \partial_{[\kappa}B_{\nu\lambda]} = 0\,,$$
$$\partial_\nu \mathfrak{H}^{\lambda\nu} = \frac{\partial}{\partial t}\mathfrak{D}^\lambda + \mathfrak{c}^\lambda\,, \quad \partial_\lambda \mathfrak{D}^\lambda = \rho\,, \quad \kappa, \nu, \lambda = 1, 2, 3\,. \tag{3.48}$$

The equations (3.48) can be used for any curvilinear coordinate system, provided one uses for the fields E_λ, $B_{\nu\lambda}$, etc., the transformations following from the already established behavior of $F_{\nu\lambda}$ and $\mathfrak{G}^{\lambda\nu}$. This means, as indicated by the notation, that E_λ and $B_{\lambda\nu}$ transform as covariant vector and bivector, while $\mathfrak{H}^{\lambda\nu}$, $\mathfrak{D}^\lambda$ and $\mathfrak{c}^\lambda$ are contravariant bivector and vector densities of weight $+ 1$ in three-space. The charge density ρ transforms as a scalar density of weight $+ 1$. It will be clear that the natural invariance of (3.48) simply follows from the natural invariance (3.11) and (3.12), because if (3.11) and (3.12) are invariant for general space-time transformations then (3.48) must be invariant for the subset of general coordinate transformations in space.

The correlation of the form (3.48) with the form (3.6) is apparently given by

$$\begin{aligned} &\boldsymbol{E} \to E_\lambda && \boldsymbol{B} \to B_{\lambda\nu} \\ &\boldsymbol{D} \to \mathfrak{D}^\lambda && \boldsymbol{H} \to \mathfrak{H}^{\lambda\nu} \\ &\boldsymbol{s} \to \mathfrak{c}^\lambda && \rho \to \rho\,. \end{aligned} \tag{3.49}$$

The identifications (3.49) can be regarded as actual equalities only for an orthogonal Cartesian frame of reference in the sense that $\boldsymbol{E} = (E_1, E_2, E_3)$ and $\boldsymbol{B} = (B_{23}, B_{31}, B_{12})$ etc.

It is interesting to note that MAXWELL [1892] (p. 13) distinguished between force vectors and flux vectors and between vectors with longitudinal and rotational symmetry. Voigt introduced the names polar and axial vectors for the latter distinction. The pairwise combination of these vector characteristics represents all the typical features of the four electromagnetic field vectors as shown in the diagram

Space vectors	Polar	Axial
force	***E***	***H***
flux	***D***	***B***

Maxwell indicated that force vectors were to be associated with line integrals and flux vectors with surface integrals.

It is possible to use the equations (3.48) to obtain the wave equations for an arbitrary stationary medium on general coordinates. Some relevant examples have been worked out in chapter 7.

We may consider a further point in checking the physical implications of Minkowski's four-dimensional form of the Maxwell equations. As shown, the Maxwell equations can be written in a four-dimensional invariant manner, provided the spatial field-vectors are combined into four-dimensional skew symmetric tensors as specified in table 2. The transformation formulae (3.14) and (3.24) impose a specific behavior of ***E***, ***B*** and ***D***, ***H*** for mutually translating frames of reference. The results of these transformations and their physical implication should conform with physical observation and can be compared with conclusions, which one can conjecture from the conventional form (3.6) of the Maxwell equations.

The simple example of a moving charge immediately suggests that there must be a "translational" relationship between ***E*** and ***B***. There is nothing much relativistic about this picture of a (slowly) moving charge and therefore, the question arises whether some first order conclusion can be obtained from fundamental relations based on plane Galilean kinematics. The integral formulation of the law of Faraday-Maxwell gives us such a possibility. This law states that

$$\oint_C \boldsymbol{E}' \cdot \mathrm{d}\boldsymbol{r} = -\frac{\mathrm{d}}{\mathrm{d}t} \int_\sigma \boldsymbol{B} \cdot \mathrm{d}\boldsymbol{\sigma} \,. \tag{3.50}$$

The time differentiation in the right-hand member implies that the loop-potential $\oint \boldsymbol{E}' \cdot \mathrm{d}\boldsymbol{r}$ can be generated by intrinsic changes in $\boldsymbol{B}$, i.e., $\partial \boldsymbol{B}/\partial t$, or/and by a motion of the loop in a nonuniform field $\boldsymbol{B}$. In the latter case the measured $\boldsymbol{E}'$ value represents an observation on a loop C which is in motion and is therefore denoted by a primed symbol.

Performing the differentiation in the right-hand member yields the known result (see § 4.2)

$$\int_\sigma \left\{ \frac{\partial \boldsymbol{B}}{\partial t} - \operatorname{curl} \boldsymbol{v} \wedge \boldsymbol{B} + \boldsymbol{v} \operatorname{div} \boldsymbol{B} \right\} \cdot \mathrm{d}\boldsymbol{\sigma} \,, \tag{3.51}$$

in which $\boldsymbol{v}$ represents the velocity of the surface $\boldsymbol{\sigma}$ associated with the contour C with respect to the frame of reference of $\boldsymbol{B}$. Substituting (3.51) in (3.50) and keeping in mind that $\operatorname{div} \boldsymbol{B} = 0$, we obtain after applying Stoke's law to the contour integral

$$\int_\sigma \operatorname{curl} (\boldsymbol{E}' - \boldsymbol{v} \wedge \boldsymbol{B}) \cdot \mathrm{d}\boldsymbol{\sigma} = - \int_\sigma \frac{\partial \boldsymbol{B}}{\partial t} \cdot \mathrm{d}\boldsymbol{\sigma} \,. \tag{3.52}$$

The integral expression (3.52) invites the definition of

$$\boldsymbol{E} = \boldsymbol{E}' - \boldsymbol{v} \wedge \boldsymbol{B} \tag{3.53}$$

as the electric field that would be observed on a stationary contour, thus leading to the conventional equation

$$\operatorname{curl} \boldsymbol{E} = -\frac{\partial \boldsymbol{B}}{\partial t}, \tag{3.54}$$

with $\boldsymbol{E}$ and $\boldsymbol{B}$ referred to the same stationary frame of reference.

A similar result can be conjectured from the Biot-Savart integral law (completed with the displacement current) leading to a transformation relation for $\boldsymbol{D}$ and $\boldsymbol{H}$

$$\boldsymbol{H} = \boldsymbol{H}' + \boldsymbol{v} \wedge \boldsymbol{D}. \tag{3.54a}$$

The equation (3.53) and (3.54) are the only ones that can be obtained about the translational transformation properties of the four field vectors $\boldsymbol{E}$, $\boldsymbol{D}$, $\boldsymbol{B}$ and $\boldsymbol{H}$ of the equations (3.6). The field equations in the form (3.6) do not give an indication that $\boldsymbol{D}$ and $\boldsymbol{B}$ transform. Therefore, the complete set for Galilean kinematics would be

$$\left.\begin{aligned} \boldsymbol{E}' &= \boldsymbol{E} + \boldsymbol{v} \wedge \boldsymbol{B} \\ \boldsymbol{H}' &= \boldsymbol{H} - \boldsymbol{v} \wedge \boldsymbol{D} \\ \boldsymbol{D}' &= \boldsymbol{D} \\ \boldsymbol{B}' &= \boldsymbol{B} \end{aligned}\right\} \text{Galilean kinematics}. \tag{3.55}$$

It is possible to obtain a first order form of Lorentz kinematics by making the observation that the equations (3.55) (which should hold in any medium) are not compatible with the empty space constitutive equations (3.7)

$$\boldsymbol{D} = \underset{0}{\varepsilon}\boldsymbol{E} \quad \text{and} \quad \boldsymbol{B} = \underset{0}{\mu}\boldsymbol{H}.$$

Substituting these into the first two equations of (3.55) yields

$$\boldsymbol{D}' = \boldsymbol{D} + \underset{0\,0}{\varepsilon\mu}\boldsymbol{v} \wedge \boldsymbol{H}$$

and

$$\boldsymbol{B}' = \boldsymbol{B} - \underset{0\,0}{\varepsilon\mu}\boldsymbol{v} \wedge \boldsymbol{E}\,,$$

instead of $\boldsymbol{D}' = \boldsymbol{D}$ and $\boldsymbol{B}' = \boldsymbol{B}$. The complete set for a first order Lorentz covariance therefore should be

$$\left.\begin{aligned} \boldsymbol{E}' &= \boldsymbol{E} + \boldsymbol{v} \wedge \boldsymbol{B} \\ \boldsymbol{H}' &= \boldsymbol{H} - \boldsymbol{v} \wedge \boldsymbol{D} \\ \boldsymbol{D}' &= \boldsymbol{D} + \underset{0\,0}{\varepsilon\mu}\boldsymbol{v} \wedge \boldsymbol{H} \\ \boldsymbol{B}' &= \boldsymbol{B} - \underset{0\,0}{\varepsilon\mu}\boldsymbol{v} \wedge \boldsymbol{E} \end{aligned}\right\} \begin{array}{l}\text{first order}\\ \text{Lorentz kinematics}\,.\end{array} \tag{3.56}$$

The reader is invited to derive exactly the equations (3.55) and (3.56) by using the equations (3.14) and (3.24) for the four-dimensional fields $\mathfrak{G}^{\lambda\nu}$ and $F_{\lambda\nu}$ and the identification C of table 2. The transformation matrices for a Galilean translation are

$$\begin{pmatrix} A_0^{0'} & A_1^{0'} & A_2^{0'} & A_3^{0'} \\ A_0^{1'} & A_1^{1'} & A_2^{1'} & A_3^{1'} \\ A_0^{2'} & A_1^{2'} & A_2^{2'} & A_3^{2'} \\ A_0^{3'} & A_1^{3'} & A_2^{3'} & A_3^{3'} \end{pmatrix} = \begin{pmatrix} 1 & 0 & 0 & 0 \\ v_1 & 1 & 0 & 0 \\ v_2 & 0 & 1 & 0 \\ v_3 & 0 & 0 & 1 \end{pmatrix}, \tag{3.57}$$

and

$$\begin{pmatrix} A_{0'}^{0} & A_{1'}^{0} & A_{2'}^{0} & A_{3'}^{0} \\ A_{0'}^{1} & A_{1'}^{1} & A_{2'}^{1} & A_{3'}^{1} \\ A_{0'}^{2} & A_{1'}^{2} & A_{2'}^{2} & A_{3'}^{2} \\ A_{0'}^{3} & A_{1'}^{3} & A_{2'}^{3} & A_{3'}^{3} \end{pmatrix} = \begin{pmatrix} 1 & 0 & 0 & 0 \\ -v_1 & 1 & 0 & 0 \\ -v_2 & 0 & 1 & 0 \\ -v_3 & 0 & 0 & 1 \end{pmatrix}. \tag{3.58}$$

The transformation matrices of a first order Lorentz translation are (MØLLER [1952] p. 117)

$$\begin{pmatrix} A_0^{0'} & A_1^{0'} & A_2^{0'} & A_3^{0'} \\ A_0^{1'} & A_1^{1'} & A_2^{1'} & A_3^{1'} \\ A_0^{2'} & A_1^{2'} & A_2^{2'} & A_3^{2'} \\ A_0^{3'} & A_1^{3'} & A_2^{3'} & A_3^{3'} \end{pmatrix} = \begin{pmatrix} 1 & \underset{0}{\varepsilon}\underset{0}{\mu} v_1 & \underset{0}{\varepsilon}\underset{0}{\mu} v_2 & \underset{0}{\varepsilon}\underset{0}{\mu} v_3 \\ v_1 & 1 & 0 & 0 \\ v_2 & 0 & 1 & 0 \\ v_3 & 0 & 0 & 1 \end{pmatrix}, \tag{3.59}$$

and

$$\begin{pmatrix} A_{0'}^{0} & A_{1'}^{0} & A_{2'}^{0} & A_{3'}^{0} \\ A_{0'}^{1} & A_{1'}^{1} & A_{2'}^{1} & A_{3'}^{1} \\ A_{0'}^{2} & A_{1'}^{2} & A_{2'}^{2} & A_{3'}^{2} \\ A_{0'}^{3} & A_{1'}^{3} & A_{2'}^{3} & A_{3'}^{3} \end{pmatrix} = \begin{pmatrix} 1 & -\underset{0}{\varepsilon}\underset{0}{\mu} v_1 & -\underset{0}{\varepsilon}\underset{0}{\mu} v_2 & -\underset{0}{\varepsilon}\underset{0}{\mu} v_3 \\ -v_1 & 1 & 0 & 0 \\ -v_2 & 1 & 1 & 0 \\ -v_3 & 0 & 0 & 1 \end{pmatrix}. \tag{3.60}$$

In (3.57)–(3.60), v_1, v_2 and v_3 are the Cartesian components of the mutual velocity of the frames of reference.

The suggested exercise is helpful to obtain a more vivid impression about the meaning of the natural invariance of the Minkowski field equations because the Galilei transformations do not have an associated space-time metric.

CHAPTER 4

VARIATIONAL ASPECTS

1. THE ACTION PRINCIPLE

The early variational principles in physics are due to Fermat, Maupertuis, Lagrange, Euler and Hamilton. It is known for instance that the fundamental equations of particle dynamics for a wide class of problems can be obtained from a variation of a time integral of the form

$$A = \int_{t_1}^{t_2} L \, \mathrm{d}t \, . \tag{4.1}$$

The functional L in (4.1), known as the Lagrangian, can be identified as the difference between kinetic and potential energy of the system described by L. The functional L, therefore, should be regarded as a function of the coordinates of the system and of the first time derivatives of the coordinates.

The time integral (4.1) determines the physical dimension of A as that of an action $[A] = [\hbar]$, which in the sense of chapter 2 suggests that A can be regarded as a legitimate 4-dimensional (domain) scalar.

One of the very attractive features of the variational principle for discrete systems is the fact that one can obtain the equations of motion from (4.1) in a form which is valid for arbitrary curvilinear coordinates.

The usefulness of the variational procedure for discrete systems has been a very enticing factor in considering generalizations of the variational techniques where the discrete system had to be replaced by a continuous system. Electromagnetics, in particular free-space electromagnetics, is a case in point where one is forced to abandon the discrete picture. There are analogies, such as kinetic energy → magnetic energy and potential energy → electrical energy, which serve as direction-finders to establish a first order relationship between the discrete and continuous systems. A most interesting textbook presentation of variational procedures for continuous systems has been given by COURANT-HILBERT [1931]. Besides the derivation of field equations, useful information can be obtained also about boundary conditions.

It will be understood that a precise, explicit form of the Lagrangian L for a continuous system is closely related to the explicit form of the so-called constitutive equations. On the other hand one likes to obtain some general conclusions from the variational principle which are valid for a certain class of very general media, without specifying the Lagrangian into its smallest detail. To do so we must distinguish between two important cases:

I. The case where one has an instantaneous and local relationship between the fields $\boldsymbol{E}$, $\boldsymbol{D}$, $\boldsymbol{H}$ and $\boldsymbol{B}$.

II. The case where one has a non-instantaneous and non-local relationship between the fields $\boldsymbol{E}$, $\boldsymbol{D}$, $\boldsymbol{H}$ and $\boldsymbol{B}$.

A very well-known example of a noninstantaneous relation is the case of a polarization field which is subject to a certain time delay before it responds to the applied electric or magnetic field strength. Along with noninstantaneous behavior one may have the situation that the polarization at a point in the medium not only depends on the field strength at that point but as well on the field strength in its neighborhood. In point of fact, assuming noninstantaneous behavior, space-time covariance will require the acceptance of nonlocal interaction as an accompanying feature.

In chapter 6 we shall find that there is an important class of media exhibiting noninstantaneous and nonlocal behavior, which can be formally reduced to Case I by the introduction of complex field variables. The coefficients in the constitutive equations then become functions of frequency.

For practical reasons we will therefore restrict ourselves to Case I. The Lagrangian L can then be represented by a single volume integral over space†

$$L = \int_\tau \mathscr{L} \, \mathrm{d}\tau \, . \tag{4.2}$$

The functional $\mathscr{L}$, which has the dimension of an energy density $[\hbar l^{-3} t^{-1}]$, should be regarded as the locally defined Lagrangian per unit volume, which is not dependent on what exists in other points of space. It carries the name Lagrangian density. Substitution of (4.2) into (4.1) leads to a 4-dimensional integral for the action

$$A = \int_{t_1}^{t_2} \mathrm{d}t \int_\tau \mathscr{L} \, \mathrm{d}\tau \, . \tag{4.3}$$

Recalling the relations between physical dimensions and transformation behavior, as discussed in chapter 2, it is immediately suggested to consider the $\mathscr{L}$ as a density in four-space

$$[\mathscr{L}] = [\hbar] \, [l^{-3} t^{-1}] \, , \tag{4.4}$$

because the factor $[l^{-3} t^{-1}]$ [see (2.6)] is a factor which charac-

† D. van Dantzig has made an interesting introductory study of the more general Case II; Proc. Royal Ac. Amsterdam **37** (1934) 526. It is believed, however, that the physical implications of his investigation have to mature before they can be regarded as textbook matter.

terizes a density of weight + 1 in the space-time manifold. The action integral (4.3) therefore should be regarded as a (scalar) integral expression, invariant for general space-time transformation

$$A = \int_{\mathfrak{f}} \mathscr{L}\, d\mathfrak{f}\,. \tag{4.5}$$

The Lagrangian density and the 4-dimensional integration element transform according to

$$\mathscr{L}(\kappa') = |\Delta|^{-1}\, \mathscr{L}(\kappa) \tag{4.6}$$

$$d\mathfrak{f}(\kappa') = |\Delta|\, d\mathfrak{f}(\kappa)\,. \tag{4.7}$$

The notation $\mathscr{L}(\kappa')$ is meant to denote that $\mathscr{L}$ depends on the coordinate $x^{\kappa'}$, whereas $\mathscr{L}(\kappa)$ is given on the coordinates x^{κ}, etc.

Next we may assume that $\mathscr{L}$ for the case of electromagnetics depends on the four-potential A_{κ} and its first derivatives with respect to the time and space coordinates, i.e.,

$$\mathscr{L} = \mathscr{L}(A_{\lambda}, \partial_{\nu} A_{\lambda})\,. \tag{4.8}$$

In addition it will be assumed that A_{λ} transforms as a co-variant vector according to table 1 given in chapter 2.

A variation of the action integral (4.5) with respect to the field A_{λ} then leads, according to well established procedures, to

$$\delta A = \int_{\mathfrak{f}} [\mathscr{L}]^{\lambda} \delta A_{\lambda}\, d\mathfrak{f} + \int_{\mathfrak{f}} \partial_{\nu} \left(\frac{\partial \mathscr{L}}{\partial(\partial_{\nu} A_{\lambda})}\, \delta A_{\lambda} \right) d\mathfrak{f}\,. \tag{4.9}$$

The notation $[\mathscr{L}]^{\lambda}$ denotes the Euler-Lagrangian derivative:

$$[\mathscr{L}]^{\lambda} = \frac{\partial \mathscr{L}}{\partial A_{\lambda}} - \partial_{\nu} \frac{\partial \mathscr{L}}{\partial(\partial_{\nu} A_{\lambda})} \,. \tag{4.10}$$

It should be noted that the second integral in (4.9) can be reduced to a 3-dimensional integral over the hypersurface enveloping the domain of space-time integration denoted by $\mathfrak{f}$.

Amongst the possible variations of A_{λ} one likes to consider only those which do not affect the transformation properties of the four-potential, i.e.

$$\delta A_{\lambda'} = A^{\lambda}_{\lambda'} \delta A_{\lambda} \,; \tag{4.11}$$

otherwise one introduces terms alien to the physical nature of A_{λ}, which would be contradictory to the idea of the concept of a so-called "virtual" variation.

The restriction (4.11) implies that the Euler-Lagrangian derivative should transform as a contravariant vector density of weight + 1, in order to maintain the invariant meaning of the first integral in (4.9). Hence

$$[\mathscr{L}]^{\lambda'} = |\,\Delta\,|^{-1} A^{\lambda'}_{\lambda} [\mathscr{L}]^{\lambda} \,. \tag{4.12}$$

Similarly, imposing separate invariance for the second integral in (4.9), one finds that the expression

$$\mathscr{L}^{\nu\lambda} = \frac{\partial \mathscr{L}}{\partial(\partial_{\nu} A_{\lambda})} \tag{4.13}$$

should transform as a tensor density of weight + 1 and contravariant valence two. Hence

$$\mathscr{L}^{\nu'\lambda'} = |\,\Delta\,|^{-1} A^{\nu'\lambda'}_{\nu\;\lambda} \mathscr{L}^{\nu\lambda} \,. \tag{4.14}$$

The condition of separate invariance of each individual integral in (4.9) follows from the consideration that (4.9) should apply for arbitrary variations of A_λ satisfying (4.11), including those for which the second integral, as a boundary integral, vanishes.

It is possible to give an explicit proof of the transformation behavior specified in (4.12) and (4.14). We may refer to the literature for those who like to see an independent confirmation of the presently obtained conclusions (SCHOUTEN [1951]; [1954]).

The variational analysis, thus far, has been restricted to Lagrangian densities which contain only first derivatives of the potential. It is of course possible to extend the procedure to an arbitrary number of derivatives. The physical significance and even the mathematical feasibility of such a generalization, however, becomes very questionable if one imposes the condition of general covariance.

Let us consider the case that $\mathscr{L}$ has first and second derivatives of the potential. The variation of the action integral (4.5) then becomes

$$\delta A = \int [\mathscr{L}]^\lambda \delta A_\lambda \, \mathrm{d}\mathfrak{f} + \int \partial_\nu \left[\left\{ \frac{\partial \mathscr{L}}{\partial(\partial_\nu A_\lambda)} - 2\partial_\tau \frac{\partial \mathscr{L}}{\partial(\partial_\nu \partial_\tau A_\lambda)} \right\} \delta A_\lambda \right] \mathrm{d}\mathfrak{f} + \int \partial_\nu \partial_\tau \left\{ \frac{\partial \mathscr{L}}{\partial(\partial_\nu \partial_\tau A_\lambda)} \delta A_\lambda \right\} \mathrm{d}\mathfrak{f} \tag{4.15}$$

with

$$[\mathscr{L}]^\lambda = \frac{\partial \mathscr{L}}{\partial A_\lambda} - \partial_\nu \frac{\partial \mathscr{L}}{\partial(\partial_\nu A_\lambda)} + \partial_\nu \partial_\tau \frac{\partial \mathscr{L}}{\partial(\partial_\nu \partial_\tau A_\lambda)} .$$

The third integral in (4.15) represents a redundant contribution in virtue of its integrand being a double divergence. Two consecutive applications of Gauss' theorem reduces this 4-dimensional integral to a 2-dimensional integral over the "contour" of a *closed* 3-dimensional domain. In other words the "contour" shrinks to a point which causes the integral to vanish. Hence, if this should be true for arbitrary coordinates and for every admissible configuration of integration, we have obtained a sufficient argument to assume that the integrand is identically zero. Also, for arbitrary configurations, we can not assume δA_λ to vanish anywhere in $\mathfrak{f}$ nor on its boundary, thus the other factor in the integrand

$$\frac{\partial \mathscr{L}}{\partial(\partial_\nu \partial_\tau A_\lambda)} \tag{4.16}$$

should have properties such that the integrand reduces to zero. In view of the symmetry in ν, τ of the double divergence, the weakest condition for which this could happen is by imposing antisymmetry on (4.16) in ν and τ, i.e.

$$\frac{\partial \mathscr{L}}{\partial(\partial_\nu \partial_\tau A_\lambda)} = - \frac{\partial \mathscr{L}}{\partial(\partial_\tau \partial_\nu A_\lambda)}\,. \tag{4.17}$$

For a well-behaved field A_λ, equation (4.17) is satisfied if no second derivatives occur in $\mathscr{L}$. A similar argument can be used for derivatives of higher order. The present result therefore reduces the variational problem to the one already treated in equations (4.9) and (4.10).

As we will find in the next chapter, the absence of derivatives of higher order than the first is closely associated with the condition of general covariance. It certainly does not imply that Lagrangians with second or higher order derivatives cannot have an *ad hoc* significance. The Lagrangians for bending bars and plates are for instance important examples with second order coordinate derivatives. It is not possible, however, to rewrite those in any simple manner in a generally covariant form, unless one returns to the original Lagrangian for the three-dimensional continuum, which has first derivatives of the displacement field only.

After having established the unique position of the Lagrangian density with first derivatives only, we may focus attention on equation (4.9) again. In view of the fact that the second integral in (4.9) is a boundary contribution we may require that both integrals vanish independently for an extremum of the action.

The necessary and sufficient condition for the first integral in (4.9) to vanish for arbitrary δA_λ, satisfying (4.11), is

$$[\mathscr{L}]^\lambda = 0\,, \tag{4.18}$$

or according to (4.10) and using the notation (4.13)

$$\partial_\nu \mathscr{L}^{\nu\lambda} = \frac{\partial \mathscr{L}}{\partial A_\lambda}\,. \tag{4.19}$$

The equation (4.19) is known to be a generally covariant expression as proven earlier in this section. The current density is zero for nonconducting media, $\mathscr{L}$ then does not depend on A_λ. Hence for nonconducting media,

$$\partial_\nu \mathscr{L}^{\nu\lambda} = 0 \tag{4.20}$$

should be a generally covariant expression. According to § 3.2 [see formulae (3.23) to (3.28)] this is possible only if

$$\mathscr{L}^{\nu\lambda} = -\,\mathscr{L}^{\lambda\nu}\,. \tag{4.21}$$

The condition (4.21) implies according to (4.13) that

$$\frac{\partial \mathscr{L}}{\partial(\partial_\nu A_\lambda)} = -\,\frac{\partial \mathscr{L}}{\partial(\partial_\lambda A_\nu)}\,, \tag{4.22}$$

so that $\mathscr{L}$ depends on the first derivatives of A_λ, only in the combination of the curl

$$2\partial_{[\nu}A_{\lambda]} = \partial_\nu A_\lambda - \partial_\lambda A_\nu\,. \tag{4.23}$$

We will find in the next chapter that the structural restriction on the Lagrangian density following from (4.21) is exclusively associated with the transformation of A_λ as a covariant vector.

The identification with the Minkowski equations (3.11) and (3.12) is now virtually unambiguous. Equation (3.11) is simply implied by the structural condition on $\mathscr{L}$ (4.21) , . . . , (4.23), and equation (3.12) is equivalent with the Euler-Lagrangian derivative of (4.9) if we take

$$\mathfrak{G}^{\nu\lambda} = \frac{\partial \mathscr{L}}{\partial(\partial_\nu A_\lambda)} = 2\,\frac{\partial \mathscr{L}}{\partial F_{\nu\lambda}}\,. \tag{4.24}$$

The factor 2 is necessary to conform with conventional definitions [see (6.12)].

Boundary conditions are most easily examined by means of the field equations directly, because it is not always certain whether a stationary value of the action integral is meaningful for the problem under consideration. However, instead of always eliminating embarrassing boundary terms in variational problems by imposing the condition that the variation of the field vanishes at the time and space boundaries of integration, one might as well consider their meaning and their use.

Having established (4.24) one can show that the second integral in (4.9) is associated with the normal component of $\boldsymbol{D}$ and the tangential component of $\boldsymbol{H}$. One may consider, e.g., a problem with a discontinuity in $\mathscr{L}$ on a closed surface in space. The interface of the discontinuity leads to two boundary integrals in (4.9), for each side of the interface, assuming the boundary integral at infinity to be zero. The principle of stationary action, extended over all space, then seems to be compatible with the familiar condition that the tangential components of $\boldsymbol{H}$ and the normal components of $\boldsymbol{D}$ should be continuous at the interface. These conditions are known to be valid provided there are no infinite charge and current densities in the neighborhood of the interface (STRATTON [1941] p. 34).

2. VECTOR-LIKE VARIATIONS OF THE POTENTIAL

In the previous paragraph we stressed the importance of the vector-like properties of the variations by means of equation (4.11). It is possible to give a somewhat more tangible and explicit form to this condition if we associate the variation in potential with an infinitesimal displacement of the domain of integration.

Consider the line integral

$$I = \int_{c} A_\lambda \, dx^\lambda , \tag{4.25}$$

taken over an open contour c as shown in Fig. 4. This integral is a legitimate, general invariant, because of the "opposite" transformation behavior of A_λ and dx^λ. Now suppose that the

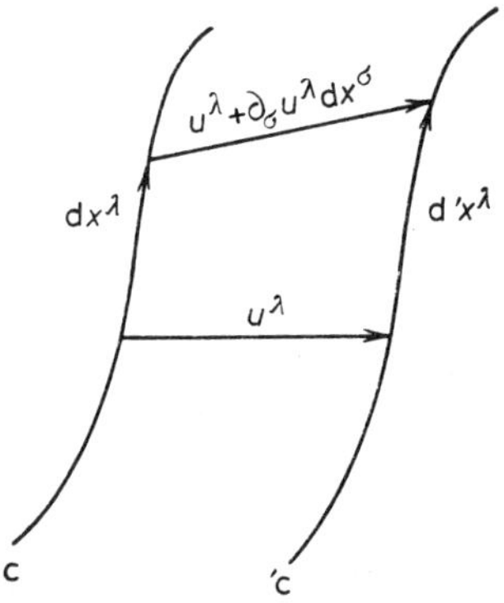

Fig. 4.

contour c goes over into 'c by the infinitesimal displacement u^λ. The vector u^λ has contravariant properties because of its very nature as a point displacement. The integral

$$'I = \int_{'c} 'A_\lambda \, d'x^\lambda , \tag{4.26}$$

over the contour 'c can be expressed as an integral over the original unprimed integration path c, by means of

$$d'x^\lambda = dx^\lambda + \partial_\sigma u^\lambda \, dx^\sigma , \tag{4.27}$$

and

$$'A_\lambda = A_\lambda + u^\sigma \partial_\sigma A_\lambda . \tag{4.28}$$

The expression (4.27) can be obtained from the quadrangle of vectors in Fig. 4, while (4.28) similarly, simply expresses the field value of $'A_\lambda$ on 'c in the field value A_λ by means of the first term of a Taylor expansion. Substituting (4.27) and (4.28) into (4.26) and discarding terms of higher order in u^σ and dx^σ one obtains for the difference

$$'I - I = \int_c (u^\sigma \partial_\sigma A_\lambda + A_\sigma \partial_\lambda u^\sigma) \, dx^\lambda \tag{4.29}$$

an integral over the original contour c. The difference $'I - I$ is a scalar, dx^λ is a contravariant vector hence the integrand

$$(u^\sigma \partial_\sigma A_\lambda + A_\sigma \partial_\lambda u^\sigma) \tag{4.30}$$

of (4.29) must be a covariant vector. The vector properties of (4.30) can be made more explicit if we rearrange its terms by partial differentiation, i.e.,

$$\{ u^\sigma 2 \partial_{[\sigma} A_{\lambda]} + \partial_\lambda (A_\sigma u^\sigma) \} .^\dagger \tag{4.30a}$$

† The 3-dimensional form in vector notation is $-\boldsymbol{u} \wedge \text{curl } \boldsymbol{A} + \text{grad } \boldsymbol{u} \cdot \boldsymbol{A}$. Following the same procedure for a surface integral of the "flux" vector $\boldsymbol{B}$ one obtains $-\text{curl } \boldsymbol{u} \wedge \boldsymbol{B} + \boldsymbol{u} \text{ div } \boldsymbol{B}$ [cf. (3.51)]. See also MADELUNG [1953] chapter VII A, § 19.

The two terms in (4.30a) are both covariant vectors according to the explanations in § 3.2, hence (4.30) is a legitimate covariant vector.

The form (4.30) [(4.30a)] is known in the mathematical literature as the Lie derivative of the field A_λ with respect to the deformation field u^λ. It will be denoted by

$$\underset{u}{\pounds} A_\lambda = u^\sigma \partial_\sigma A_\lambda + A_\sigma \partial_\lambda u^\sigma . \tag{4.31}$$

There is a similar rule as for covariant derivatives for writing the Lie derivatives of other fields†. For instance the Lie derivative of a scalar density $\mathscr{L}$ is

$$\underset{u}{\pounds} \mathscr{L} = u^\sigma \partial_\sigma \mathscr{L} + \mathscr{L} \partial_\sigma u^\sigma = \partial_\sigma (\mathscr{L} u^\sigma) . \tag{4.32}$$

It is useful to note that the Lie derivative, in contradistinction to the covariant derivative, requires only the presence of a displacement field in the manifold. The use of the Lie derivative in field problems, therefore, has no immediate implications with regard to the physical interpretation of the structure of the manifold. In other words one does not commit oneself to delicate decisions on the physical meaning of the space-time structure when using Lie derivatives.

Consider a deformation of the action integral, which according to a similar theorem as (4.29) is

$$\int \underset{u}{\pounds} \mathscr{L} \, \mathrm{d}\mathfrak{f} . \tag{4.33}$$

† SCHOUTEN [1951] and [1954]. The Lie derivative of a scalar φ is simply $u^\sigma \partial_\sigma \varphi$, because the φ does not transform. If $\varphi = A_\lambda B^\lambda$ one finds that $\underset{u}{\pounds}\varphi = B^\lambda \underset{u}{\pounds} A_\lambda + A_\lambda \underset{u}{\pounds} B^\lambda$ with $\underset{u}{\pounds} B^\lambda = u^\sigma \partial_\sigma B^\lambda - B^\sigma \partial_\sigma u^\lambda$ as the analogue of (4.31) for a contravariant vector. Similarly one finds the Lie derivatives of more complex tensor fields. Please note that the Leibniz rule for differentiation holds.

We may assume next that the Lagrangian density in (4.33) represents a nonconducting medium. The $\mathscr{L}$ is then a function of the first derivatives of the potentials and furthermore depends on the structural field. The structural field will be left unspecified, it will be denoted by ST. Hence

$$\mathscr{L} = \mathscr{L}(\partial_\nu A_\lambda, \mathrm{ST}) . \tag{4.34}$$

The expression (4.33) then gives rise to the identity

$$\int \underset{u}{\pounds} \mathscr{L} \, \mathrm{d}\mathfrak{f} = \int \frac{\partial \mathscr{L}}{\partial(\partial_\nu A_\lambda)} \underset{u}{\pounds}(\partial_\nu A_\lambda) \, \mathrm{d}\mathfrak{f} + \int \frac{\partial \mathscr{L}}{\partial(\mathrm{ST})} \underset{u}{\pounds}(\mathrm{ST}) \, \mathrm{d}\mathfrak{f} . \tag{4.35}$$

Inserting the explicit forms of the Lie derivatives as given in (4.31) and (4.32) in the first two integrals of (4.35), one can convert the identity into the following form after performing a few perfectly straightforward partial integrations:

$$\int_{\mathfrak{f}} \partial_\nu \{ (\mathscr{L} \delta_\lambda^\nu - \mathfrak{G}^{\nu\sigma} F_{\lambda\sigma}) u^\lambda \} \, \mathrm{d}\mathfrak{f} = \int_{\mathfrak{f}} \frac{\partial \mathscr{L}}{\partial(\mathrm{ST})} \underset{u}{\pounds}(\mathrm{ST}) \, \mathrm{d}\mathfrak{f} . \tag{4.36}$$

For the transition from (4.35) to (4.36) one makes use of the fact that the field equations are satisfied. Furthermore one uses the property

$$\underset{u}{\pounds} \partial_\nu A_\lambda = \partial_\nu \underset{u}{\pounds} A_\lambda . \tag{4.37}$$

In the course of the conversion one ends up with a double divergence integral which can be ruled out on basis of the skew symmetry (4.2). In point of fact one can use this as another argument that $\partial \mathscr{L}/\partial(\partial_\nu A_\lambda)$ should be skew symmetric in ν and λ.

The integral expression (4.36) is an important relation in the discussion of admissible Lagrangians (§ 5.3). The expression between parentheses in the integrand of the integral on the left-hand side of (4.36) will turn out to be the energy momentum tensor of the electromagnetic field. A structural field ST with a vanishing Lie derivative for arbitrary deformation u^λ thus must lead to an energy momentum tensor that is identically zero.

3. INTERPRETATION OF ENERGY AND STRESS IN AN ARBITRARY (NONDISPERSIVE) MEDIUM

The energy and stress relations can be immediately derived from the Minkowski equations (3.11) and (3.12). Multiplying (3.12) with $F_{\sigma\lambda}$ and summing over λ one gets

$$F_{\sigma\lambda}\partial_\nu \mathfrak{G}^{\lambda\nu} = \mathfrak{c}^\lambda F_{\sigma\lambda}, \tag{4.38}$$

and rearranging the left-hand member by partial differentiation, one obtains

$$\partial_\nu(F_{\sigma\lambda}\mathfrak{G}^{\lambda\nu}) - \mathfrak{G}^{\lambda\nu}\partial_\nu F_{\sigma\lambda} = \mathfrak{c}^\lambda F_{\sigma\lambda}. \tag{4.39}$$

Remembering that according to (3.11)

$$\partial_{[\nu}F_{\sigma\lambda]} = 0,$$

or

$$\partial_\nu F_{\sigma\lambda} + \partial_\lambda F_{\nu\sigma} + \partial_\sigma F_{\lambda\nu} = 0, \tag{4.40}$$

one can use this to modify the second term in the left-hand member of (4.39). Multiplying (4.40) with $\mathfrak{G}^{\lambda\nu}$ and summing over λ, ν, one finds after using the skew symmetry of the fields

$$-\mathfrak{G}^{\lambda\nu}\partial_\nu F_{\sigma\lambda} = \tfrac{1}{2}\mathfrak{G}^{\lambda\nu}\partial_\sigma F_{\lambda\nu}. \tag{4.41}$$

Substitution of (4.41) in (4.39) yields

$$\boxed{\partial_\nu(F_{\sigma\lambda}\mathfrak{G}^{\lambda\nu}) + \tfrac{1}{2}\mathfrak{G}^{\lambda\nu}\partial_\sigma F_{\lambda\nu} = F_{\sigma\lambda}\mathfrak{c}^\lambda} \tag{4.42}$$

So far all the operations that have been performed to obtain (4.42) are legitimate invariant operations. The right-hand member of (4.42) represents a covariant vector density (Lorentz force density), hence the left-hand side should represent a covariant vector density. It should be understood that it is the sum of the two left-hand members which represents a covariant vector density, not the two terms individually. One can prove this statement by the technique demonstrated in § 3.2, though an explicit proof is hardly necessary because the structure of equation (4.42) reveals everything we need to know. The customary way of obtaining the stress energy relations from (4.42) is to assume that the medium is linear, i.e., the field $\mathfrak{G}^{\lambda\nu}$ is component-wise proportional to $F_{\lambda\nu}$. Under those circumstances one can write for the second term in the left-hand member of (4.42)

$$\tfrac{1}{2}\mathfrak{G}^{\lambda\nu}\partial_\sigma F_{\lambda\nu} = \tfrac{1}{4}\partial_\sigma(\mathfrak{G}^{\lambda\nu}F_{\lambda\nu})\,, \tag{4.43}$$

thus reducing (4.42) to the familiar form

$$\partial_\nu(\tfrac{1}{4}\delta^\nu_\sigma\mathfrak{G}^{\lambda\kappa}F_{\lambda\kappa} - F_{\lambda\sigma}\mathfrak{G}^{\lambda\nu}) = F_{\sigma\lambda}\mathfrak{c}^\lambda\,. \tag{4.44}$$

It is probably not needless to draw attention to the fact that the left-hand member of (4.44) is not a covariant vector density for general coordinates. One can check this again by means of the technique of § 3.2. The expression (4.44) is valid only for linear media on a Lorentz frame of reference with Cartesian space coordinates.

Having summarized the limitations of the customary transition (4.42) to (4.44), we may look for a conversion of (4.42) which satisfies a wider range of possibilities. Suppose that the problem under discussion can be expressed by a variational principle with a Lagrangian density $\mathscr{L}$. Using our present knowledge about the structure of $\mathscr{L}$ as substantiated by the relations (4.17) and (4.22), we can write down the following identity

$$\partial_\sigma \mathscr{L} = \partial_{(\sigma)} \mathscr{L} + \frac{\partial \mathscr{L}}{\partial(\partial_\lambda A_\nu)} \partial_\sigma \partial_{[\lambda} A_{\nu]} + \frac{\partial \mathscr{L}}{\partial A_\nu} \partial_\sigma A_\nu \, , \qquad (4.45)$$

in which $\partial_{(\sigma)} \mathscr{L}$ denotes the explicit derivative of $\mathscr{L}$ with respect to x^σ, i.e., A_ν and $\partial_\lambda A_\nu$ regarded as constants.

For a medium which carries charge and current one may expect the Lagrangian density to be composed of two additive parts

$$\mathscr{L} = \underset{0}{\mathscr{L}} + \underset{1}{\mathscr{L}} \, , \qquad (4.46)$$

as suggested by the field equation (4.19), i.e., $\underset{0}{\mathscr{L}}$ depends on the derivatives of the potentials only and $\underset{1}{\mathscr{L}}$ depends on the potentials itself;

$$\begin{aligned} \underset{0}{\mathscr{L}} &= \underset{0}{\mathscr{L}} (\partial_\nu A_\lambda) \, , \\ \underset{1}{\mathscr{L}} &= \underset{1}{\mathscr{L}} (A_\nu) \, . \end{aligned} \qquad (4.47)$$

Using (4.24), (4.46) and (4.47) in (4.45) one finds

$$\frac{\partial \mathscr{L}}{\partial(\partial_\lambda A_\nu)} \partial_\sigma \partial_{[\lambda} A_{\nu]} = \tfrac{1}{2} \mathfrak{G}^{\lambda\nu} \partial_\sigma F_{\lambda\nu} = \partial_\sigma \underset{0}{\mathscr{L}} - \partial_{(\sigma)} \underset{0}{\mathscr{L}} \, . \qquad (4.48)$$

By means of (4.48), the relation (4.42) can now be written as

$$\boxed{\partial_\nu \{ \delta^\nu_\sigma \underset{0}{\mathscr{L}} - F_{\lambda\sigma} \mathfrak{G}^{\lambda\nu} \} - \partial_{(\sigma)} \underset{0}{\mathscr{L}} = F_{\sigma\lambda} \mathfrak{c}^\lambda} \qquad (4.49)$$

The equation (4.49) is a generally invariant expression, provided $\underset{0}{\mathscr{L}}$ satisfies the restrictions (4.17) and (4.22) as well as another relation depending on the structural fields in $\underset{0}{\mathscr{L}}$. These restrictions, as will be discussed in the next chapter, are the compatibility relations for so-called admissible Lagrangian structures. A direct proof of the general covariance of (4.49) is possible if one has an explicit form of $\underset{0}{\mathscr{L}}$. An important example of such an explicit form will be discussed in chapter 9.

In the same way as for (4.42), it should be kept in mind that it is the sum, or rather the difference, of the two left-hand members that transforms as a covariant vector density. The two terms individually don't have tensor properties for general coordinates. One may compare this situation with the equation of the geodesic

$$\ddot{x}^{\kappa} + \Gamma^{\kappa}_{\lambda\nu}\dot{x}^{\lambda}\dot{x}^{\nu} = 0 \,.$$

The first term $\ddot{x}^{\kappa}$ nor the second term $\Gamma^{\kappa}_{\lambda\nu}\dot{x}^{\lambda}\dot{x}^{\nu}$ are tensors (contravariant vectors); their sum, however, is a legitimate contravariant vector for general coordinates. It is very necessary to keep these things in mind for a physical interpretation of the two left-hand terms of (4.49) in order to keep coordinates and physical interpretation separated.

A simple case for eliminating coordinate reduncancy is given when we can assume a linear frame of reference. One thus discards the nonuniformity of space-time, considering the fact that an actual material medium can be expected to introduce nonuniformities of much larger magnitude than the inhomogeneity of space-time itself. Under those conditions one may associate the following interpretations with the three terms of equation (4.49):

I	$\mathfrak{c}^{\lambda}F_{\sigma\lambda}$	$\sigma = 1, 2, 3$	Force interaction with current and charge.
		$\sigma = 0$	Energy interaction with current and charge.

II	$\partial_{(\sigma)}\underset{0}{\mathscr{L}}$	$\sigma = 1, 2, 3$	Force interaction with the medium.
		$\sigma = 0$	Energy interaction with the medium.
III	$\partial_\nu\{\delta^\nu_\sigma\underset{0}{\mathscr{L}} - F_{\lambda\sigma}\mathfrak{G}^{\lambda\nu}\}$	$\sigma = 1, 2, 3$	Force interaction with the field.
		$\sigma = 0$	Energy interaction with the field.

We will submit the identifications I, II and III to a point-wise discussion:

(I) Using the identification C of table 2, chapter 3, one easily shows that the term I corresponds to the Lorentz force and the rate of change of energy density per unit time

$$\mathfrak{c}^\lambda F_{\sigma\lambda}\begin{cases}\rho\boldsymbol{E} + \boldsymbol{s}\wedge\boldsymbol{B} & \sigma = 1, 2, 3\\ \boldsymbol{s}\cdot\boldsymbol{E} & \sigma = 0\,.\end{cases} \tag{4.50}$$

We may refer to standard texts for the physical interpretation of these familiar expressions. Please note that ρ and $\boldsymbol{s}$ represent the macroscopic charge and current density.

(II) This term vanishes if the properties of the medium are uniform throughout space and independent of the time, because the explicit derivative by definition operates only on the structural parameters of $\underset{0}{\mathscr{L}}$, e.g., permittivity and permeability.

Suppose one has a linear isotropic, nonuniform and non-conducting medium with permeability μ and permittivity ε. The Lagrangian density is then given by the expression

$$\underset{0}{\mathscr{L}} = \frac{1}{2}\frac{1}{\mu}\boldsymbol{B}^2 - \tfrac{1}{2}\varepsilon\boldsymbol{E}^2\,. \tag{4.51}$$

The force per unit volume acting on the medium is

$$\partial_{(\sigma)}\underset{0}{\mathscr{L}} = -\left(\frac{1}{2}\frac{\boldsymbol{B}^2}{\mu^2}\frac{\partial\mu}{\partial x^\sigma} + \tfrac{1}{2}\boldsymbol{E}^2\frac{\partial\varepsilon}{\partial x^\sigma}\right)$$

or

$$-\partial_{(\sigma)}\underset{0}{\mathscr{L}} = \frac{1}{2}\left(\boldsymbol{E}^2\frac{\partial\varepsilon}{\partial x^\sigma} + \boldsymbol{H}^2\frac{\partial\mu}{\partial x^\sigma}\right). \tag{4.52}$$

Now assume that the nonuniformity occurs on a plane surface only, where ε and μ exhibit finite step discontinuities; they are ε_1, μ_1 in the region 1 and ε_2, μ_2 in the region 2 at either side of the interface. Next we suppose that an electromagnetic wave hits the interface under normal incidence. The normal is taken as the direction of the x coordinate.

The force density (4.52) becomes infinite at the interface, but the force f_n per unit area is the finite limit

$$f_\mathrm{n} = \lim_{\Delta x\to 0} \Delta x \frac{\Delta\mathscr{L}_0}{\Delta x}, \tag{4.53}$$

which according to (4.52) leads to the familiar result

$$f_\mathrm{n} = \tfrac{1}{2}(\boldsymbol{E}^2\varepsilon_2 + \boldsymbol{H}^2\mu_2) - \tfrac{1}{2}(\boldsymbol{E}^2\varepsilon_1 + \boldsymbol{H}^2\mu_1). \tag{4.54}$$

The transition from (4.52) to (4.54) is allowed, because the fields **E** and **H** which are tangential to the surface (normal incidence) are continuous across the interface.

The conclusion given by (4.54) is that the radiation force f_n on the interface is equal to the difference of the energy densities at either side of the discontinuity. A quick look at (4.49) shows that the same conclusion can be obtained from the first term in the left-hand member of this equation, because the right-hand side is zero for a nonconducting charge-free medium.

Having identified the space components of $\partial_{(\sigma)} \underset{0}{\mathscr{L}}$ in (4.49) as the force components acting on the medium per unit volume, one finds that the time component ($\sigma = 0$) represents the energy sinks or energy sources of the medium.

(III) The expression in curly brackets listed under III is a new form of the energy momentum tensor which was already briefly mentioned at the end of the previous section. First of all we may remark that it reduces to the customary expressions given by (4.44) if the medium is linear and uniform in space and time, i.e., $\partial_{(\sigma)} \underset{0}{\mathscr{L}} = 0$. But let us see in detail in what respect the general case deviates from the customary expressions for energy and momentum of the electromagnetic field†. A component-wise representation of the tensor

$$\mathfrak{T}_{\sigma}^{\cdot\nu} = \{\delta_{\sigma}^{\nu} \underset{0}{\mathscr{L}} - F_{\sigma\lambda} \mathfrak{G}^{\nu\lambda}\} \tag{4.55}$$

leads to the following matrix

$$\begin{pmatrix} \underset{0}{\mathscr{L}} - F_{0\lambda}\mathfrak{G}^{0\lambda} & -F_{0\lambda}\mathfrak{G}^{1\lambda} & -F_{0\lambda}\mathfrak{G}^{2\lambda} & -F_{0\lambda}\mathfrak{G}^{3\lambda} \\ -F_{1\lambda}\mathfrak{G}^{0\lambda} & \underset{0}{\mathscr{L}} - F_{1\lambda}\mathfrak{G}^{1\lambda} & -F_{1\lambda}\mathfrak{G}^{2\lambda} & -F_{1\lambda}\mathfrak{G}^{3\lambda} \\ -F_{2\lambda}\mathfrak{G}^{0\lambda} & -F_{2\lambda}\mathfrak{G}^{1\lambda} & \underset{0}{\mathscr{L}} - F_{2\lambda}\mathfrak{G}^{2\lambda} & -F_{2\lambda}\mathfrak{G}^{3\lambda} \\ -F_{3\lambda}\mathfrak{G}^{0\lambda} & -F_{3\lambda}\mathfrak{G}^{1\lambda} & -F_{3\lambda}\mathfrak{G}^{2\lambda} & \underset{0}{\mathscr{L}} - F_{3\lambda}\mathfrak{G}^{3\lambda} \end{pmatrix}, \tag{4.55a}$$

in which λ takes the values 0, 1, 2 and 3.

Using table 2 of chapter 3 one can transcribe (4.55a) into the more customary three-dimensional symbols leading to the form

† As a basis of comparison we take the form used by Einstein (cf. LORENTZ *et al.* [1922] p. 118, for free space), which is identical with (4.44).

$$\begin{pmatrix} (\underset{0}{\mathcal{L}}+\mathbf{E}\cdot\mathbf{D}) & (\mathbf{E}\wedge\mathbf{H})_1 & (\mathbf{E}\wedge\mathbf{H})_2 & (\mathbf{E}\wedge\mathbf{H})_3 \\ -(\mathbf{D}\wedge\mathbf{B})_1 & (\underset{0}{\mathcal{L}}+E_1D_1-H_2B_2-H_3B_3) & (E_1D_2+H_1B_2) & (E_1D_3+H_1B_3) \\ -(\mathbf{D}\wedge\mathbf{B})_2 & (E_2D_1+H_2B_1) & (\underset{0}{\mathcal{L}}+E_2D_2-H_1B_1-H_3B_3) & (E_2D_3+H_2B_3) \\ -(\mathbf{D}\wedge\mathbf{B})_3 & (E_3D_1+H_3B_1) & (E_3D_2+H_3B_2) & (\underset{0}{\mathcal{L}}+E_3D_3-H_1B_1-H_2B_2) \end{pmatrix}. \tag{4.55b}$$

A transcription of the energy momentum tensor occurring in the form (4.44), which represents the linear (uniform) case, leads to

$$\begin{pmatrix} \mathcal{E} & (\mathbf{E}\wedge\mathbf{H})_1 & (\mathbf{E}\wedge\mathbf{H})_2 & (\mathbf{E}\wedge\mathbf{H})_3 \\ -(\mathbf{D}\wedge\mathbf{B})_1 & (\mathcal{E}-E_1D_1-H_1B_1) & (E_1D_2+H_1B_2) & (E_1D_3+H_1B_3) \\ -(\mathbf{D}\wedge\mathbf{B})_2 & (E_2D_1+H_2B_1) & (\mathcal{E}-E_2D_2-H_2B_2) & (E_2D_3+H_2B_3) \\ -(\mathbf{D}\wedge\mathbf{B})_3 & (E_3D_1+H_3B_1) & (E_3D_2+H_3B_2) & (\mathcal{E}-E_3D_3-H_3B_3) \end{pmatrix} \tag{4.44a}$$

with $\mathcal{E} = \frac{1}{2}(\mathbf{E}\cdot\mathbf{D} + \mathbf{H}\cdot\mathbf{B})$ as the energy density of the linear medium.

An inspection of the two forms (4.55b) and (4.44a) shows that there is a difference only in the diagonal terms. The expressions for the energy flow $\boldsymbol{E} \wedge \boldsymbol{H}$, the electromagnetic momentum $\boldsymbol{D} \wedge \boldsymbol{B}$ and the off-diagonal stresses $(\boldsymbol{E}_1\boldsymbol{D}_2 + \boldsymbol{H}_1\boldsymbol{B}_2)$ etc. are exactly the same.

The two matrices (4.55b) and (4.44a) are completely identical if one assumes a Lagrangian density given by

$$\underset{0}{\mathscr{L}} = \tfrac{1}{2}\boldsymbol{H}\cdot\boldsymbol{B} - \tfrac{1}{2}\boldsymbol{E}\cdot\boldsymbol{D}\,, \tag{4.56}$$

which in point of fact represents the linear medium. But please note that linearity does not imply the identity of (4.49) and (4.44); their identity in addition requires uniformity, i.e., $\partial_{(\sigma)}\underset{0}{\mathscr{L}} = 0$.

For a general medium for which one does not have an explicit $\underset{0}{\mathscr{L}}$, equation (4.24) can be used to obtain the expression

$$\mathrm{d}\underset{0}{\mathscr{L}} = \tfrac{1}{2}\mathfrak{G}^{\lambda\nu}\,\mathrm{d}F_{\lambda\nu}$$

or

$$\mathrm{d}\underset{0}{\mathscr{L}} = \boldsymbol{H}\cdot\mathrm{d}\boldsymbol{B} - \boldsymbol{D}\cdot\mathrm{d}\boldsymbol{E}\,, \tag{4.57}$$

for the differential of $\underset{0}{\mathscr{L}}$.

The $\underset{0}{\mathscr{L}}$ itself can then be represented by the integral expressions

$$\underset{0}{\mathscr{L}} = \int\limits_{B=0}^{B} \boldsymbol{H}\cdot\mathrm{d}\boldsymbol{B} - \int\limits_{E=0}^{E} \boldsymbol{D}\cdot\mathrm{d}\boldsymbol{E}\,, \tag{4.58}$$

with $E = 0$ and $B = 0$ as the reference states for vanishing $\underset{0}{\mathscr{L}}$.

Now, assuming that the component $\mathfrak{T}_0^{\cdot 0}$ of (4.55) represents the density of electromagnetic energy, one obtains

$$\mathfrak{T}_0^{\cdot 0} = \int\limits_{B=0}^{B} \boldsymbol{H}\cdot\mathrm{d}\boldsymbol{B} - \int\limits_{E=0}^{E} \boldsymbol{D}\cdot\mathrm{d}\boldsymbol{E} + \boldsymbol{E}\cdot\boldsymbol{D}$$

or

$$\mathfrak{T}_{0}^{\cdot 0} = \int_{B=0}^{B} \boldsymbol{H} \cdot \mathrm{d}\boldsymbol{B} + \int_{E=0}^{E} \boldsymbol{E} \cdot \mathrm{d}\boldsymbol{D}\,. \tag{4.59}$$

The expression (4.59) is in accordance with the customary forms that have been adopted for the energy content of non-linear and multivalued (hysteresis) media, except that for the dielectric energy, $\boldsymbol{E} = 0$ instead of $\boldsymbol{D} = 0$ has been chosen as the reference state.

The assumption of $\boldsymbol{B} = 0$ as a reference state has been extensively discussed in the literature of ferromagnetics. This being a formal presentation of the subject matter, it may suffice to mention that $\boldsymbol{B} = 0$ necessarily requires $\boldsymbol{E} = 0$ on ground of the fact that the reference state should not be affected by a Lorentz transformation. This reference state, incidentally, is the natural state for a superconductor.

Summarizing the situation after the point-wise discussion of the three terms of (4.49), we may conclude that this relation indeed gives a satisfactory account of energy and momentum exchange in electromagnetics.

The energy momentum relations in electromagnetics have been a point of concern for a long time. HERTZ [1890] raised the question that the customary relations do not really account for an interaction with a nonconducting charge-free medium. Lorentz† took up the problem much later and came up with a balance of stresses rather than a balance of forces as given by (4.49).

The example, worked out in § 4.2, of the radiation forces on an interface between two media, and also the fact that (4.55) automatically leads to a meaningful energy expression for nonlinear media, are strong arguments to accept (4.49) as a

† H. A. Lorentz, Encyclopädie der Mathematischen Wissenschaften, Volume V, Part 2 (Teubner, Leipzig, 1904–1922), p. 107.

significant improvement over the customary formulations.

One should keep in mind, however, that applications of (4.49) depend critically on the structure of the Lagrangian density. The omission of structural fields that are essential for a given physical situation, or the incorporation of structural fields in the wrong manner, will certainly lead to erroneous results. The compatibility check for the Lagrangians (see chapter 5) should therefore be considered as an important means to guard against faulty structures.

It may be worthwhile to draw attention to the analogous problem in acoustics which likewise, but for slightly different reasons, has been the subject of long standing controversy. In acoustics it is possible, again from a variational principle, to deduce a relation similar to (4.49) for what may be formally called the interaction of field and medium, POST [1960]. The explicit derivative $\partial_{(\sigma)} \underset{0}{\mathscr{L}}$ again represents the force on the medium (the analogue of force on current and charges is of course absent). The author is indebted to G. Weinreich for pointing out the importance of these relations for electromagnetics and for acoustics.

In connection with the previous remark about the similarity to acoustics it may be of some interest to realize that the time average of the Maxwell stress, for a single frequency traveling wave in a linear medium, can also be interpreted as the flux of the electromagnetic momentum $\boldsymbol{D} \wedge \boldsymbol{B}$. The reader is invited to prove for this purpose the relation

$$\langle \mathscr{E}\boldsymbol{I} - \boldsymbol{ED} - \boldsymbol{HB} \rangle = \frac{1}{\mathscr{E}} \langle (\boldsymbol{D} \wedge \boldsymbol{B})(\boldsymbol{E} \wedge \boldsymbol{H}) \rangle = \frac{\mathscr{E}}{\omega} \boldsymbol{kg}\,. \tag{4.60}$$

The vector $\boldsymbol{g}$ is the group velocity of the wave. Please note that the group velocity and phase velocity are already different in a nondispersive, anisotropic medium. The juxtaposition of vectors refers to Gibb's dyadic product, i.e., $\boldsymbol{ED}$

represents the tensor product $E_a D^b$ and I is the unit tensor.

The corresponding energy momentum tensor for a photon-beam can accordingly be given the very simple form

$$\hbar k_\lambda \mathfrak{N}^\nu , \tag{4.61}$$

with $\hbar k_\lambda$ the momentum per photon and $\mathfrak{N}^\nu$ the vector density of the photon stream.

It should be kept in mind that the explanations in this chapter and section are based on the assumption of local and instantaneous section of the fields which in principle excludes the treatment of dispersive media. As mentioned in § 4.1, the introduction of complex fields provides a partial escape from this restriction†.

† The author acknowledges the kind permission for the use of unpublished work by G. Weinreich for the preparation of this chapter.

CHAPTER 5

LAGRANGIANS

1. ADMISSIBLE LAGRANGIAN DENSITIES

In the previous chapter we discussed at some length restrictions on the structure of Lagrangian densities as expressed by the formulae (4.17) and (4.22). We may attempt to obtain a somewhat clearer idea of where these restrictions come from and what they mean. Before considering details it is necessary to answer questions about the kind of fields one may expect as constituents of Lagrangians and last, not least, their transformation behavior.

First of all, let us distinguish between two important constructional elements which should occur in any Lagrangian. They are:

The *functional* field and its derivatives and the *structural* field. The four-potential is an example of the first one, whereas permittivity and permeability are examples of the latter. The structural field should be regarded as a thing that is given *a priori* specifying the properties of the medium. The functional field specifies the state of the medium.

Depending on the problem under consideration, it is conceivable that some field may occur as a structural or as a functional field. The metrical structure of space-time for instance, plays the role of a structural field in electromagnetic problems; for gravitational theory in the sense of general relativity it is regarded as a functional field associated with the "state" of the space-time manifold as a function of the distribution of matter.

A criterion for the distinction between functional and structural fields can be found in the so-called Neumann principle. The structural field should be subject to the Neumann principle. The functional field is independent of the restrictions imposed by this principle.

The Neumann or Brewster principle, in its restricted version as used in crystal physics, states that the physical properties as well as the outward geometrical properties of crystals are invariant with respect to certain finite groups of rotations and reflections. Any such group characterizes the symmetry of a particular crystal and determines the potential possibility for crystals to have properties such as birefringence, optical activity, piezoelectricity, etc.

It was W. VOIGT [1910] who recognized most explicitly the importance of the transformation behavior of physical properties for a meaningful application of the Neumann principle. The results of his investigations were laid down in his monumental "Lehrbuch der Kristallphysik" [1910]. Voigt is the originator of the name "tensor", though the tensor concept in principle occurred more or less simultaneously in geometry and physics around the turn of the century. Before that time tensor-like quantities of course did occur; there was, however, much less emphasis on their typical mathematical characteristics.

The Neumann principle in its restricted form is usually associated with the 32 finite groups of rotations and reflections in crystal physics. It is important, however, to recognize the Neumann principle as a general principle of physics, applicable to arbitrary media. The Lorentz group, for instance, can be regarded as the group characterizing properties of free space. The invariance against time reversal will prove a powerful criterion to distinguish between reciprocal and nonreciprocal media.

Summarizing the situation we can say that the constituent

fields of a Lagrangian density are of two kinds. One of them has been called the functional field; it determines the state of the medium. The Lagrangian depends on this field and on its derivatives. The other field, which we called the structural field, informs us about the properties of the medium. It is the structural field only that is subject to the Neumann principle in its most general form.

Having established the nature of the two constituent fields, the problem of admissible Lagrangians becomes the problem of constructing differential concomitants of these constituent fields so that the resulting Lagrangian has the transformation properties of a four-dimensional density as given by (4.6).

The important cases for electromagnetics are contained in the choice:

A. The four-potential A_λ and its derivatives represent the functional field.

B. The structural field will be assumed to be a tensor field.

For the time being we can leave the structural field unspecified, except that it obeys the homogeneous transformation law of a tensor. Specific cases will be discussed later. We may denote the unspecified structural tensor field by ST as in (4.34). The transformation rule (4.6) of the Lagrangian density then can be written in the form

$$\mathscr{L}(A_{\lambda'}, \partial_{\nu'} A_{\lambda'}, \mathrm{ST}') = |\varDelta|^{-1} \mathscr{L}(A_\lambda, \partial_\nu A_\lambda, \mathrm{ST}) \tag{5.1}$$

with

$$A_{\lambda'} = A^{\lambda}_{\lambda'} A_\lambda , \tag{5.2}$$

$$\partial_{\nu'} A_{\lambda'} = A^{\nu\lambda}_{\nu'\lambda'} \partial_\nu A_\lambda + A_\lambda \partial_{\nu'} A^{\lambda}_{\lambda'} , \tag{5.3}$$

plus the information that the transformation

$$\mathrm{ST}' \rightarrow \mathrm{ST} , \tag{5.4}$$

invokes only the coefficients $A^{\lambda}_{\lambda'}$ but not their derivatives $\partial_{\nu'} A^{\lambda}_{\lambda'}$.

An inspection of the equations (5.1) ... (5.4) then shows that the transformation (5.1) can hold only if the derivatives of $A^{\lambda}_{\lambda'}$ cancel out in the structure of $\mathscr{L}$. This condition can be met if the derivatives of A_λ occur in $\mathscr{L}$ in the form of the curl of A_λ only. Proof: equation (5.1) should be an identity in the transformation elements $A^{\lambda}_{\lambda'}$ and in the derivatives $\partial_{\nu'} A^{\lambda}_{\lambda'}$ of the transformation elements. The latter condition applied to (5.1) gives

$$\frac{\partial \mathscr{L}}{\partial(\partial_{\nu'} A^{\lambda}_{\lambda'})} \equiv 0 \tag{5.5}$$

or

$$\frac{\partial \mathscr{L}}{\partial(\partial_{\nu'} A^{\lambda}_{\lambda'})} = \frac{\partial \mathscr{L}}{\partial(\partial_{\sigma'} A_{\kappa'})} \frac{\partial(\partial_{\sigma'} A_{\kappa'})}{\partial(\partial_{\nu'} A^{\lambda}_{\lambda'})} \equiv 0\,, \tag{5.6}$$

because comparing (1.10), (5.2), (5.3) and (5.4), it is only (5.3) which invokes the derivatives of the elements $A^{\lambda}_{\lambda'}$. The last factor in the second member of (5.6) can be converted according to (5.3) into

$$\frac{\partial(\partial_{\sigma'} A_{\kappa'})}{\partial(\partial_{\nu'} A^{\lambda}_{\lambda'})} = \delta^{\nu'}_{\sigma'} \delta^{\lambda'}_{\kappa'} A_\lambda\,. \tag{5.7}$$

For a holonomic transformation (1.13), (5.7) can be written in the form

$$\frac{\partial(\partial_{\sigma'} A_{\kappa'})}{\partial(\partial_{\nu'} A^{\lambda}_{\lambda'})} = \delta^{(\nu'}_{\sigma'} \delta^{\lambda')}_{\kappa'} A_\lambda\,. \tag{5.7a}$$

Hence substitution of (5.7a) into (5.6) yields (4.22).

Similar but more cumbersome is the proof that higher order derivatives cannot be admitted in $\mathscr{L}$. One imposes the condition that (5.1) should be an identity in $\partial_{\tau'} \partial_{\nu'} A^{\lambda}_{\lambda'}$, and one obtains the result expressed by (4.17). A somewhat more subtle argument is required to show that the statement is true for derivatives of arbitrary order higher than the first.

2. TRIVIAL LAGRANGIANS AND THE PFAFFIAN CLASSIFICATION OF THE POTENTIALS

It is a known fact that the covariant four-potential as a rule, cannot be represented by a gradient of a scalar, i.e.,

$$A_\lambda = \partial_\lambda \Psi \, , \tag{5.8}$$

because then the field

$$F_{\nu\lambda} = 2\partial_{[\nu} A_{\lambda]} \tag{5.9}$$

would be identically zero, which incidentally is true only for the superconducting state.

The next simplest possibility is to regard A_λ as a vector field parallel to a gradient field, i.e.,

$$A_\lambda = \underset{1}{\psi} \partial_\lambda \underset{2}{\psi} \, . \tag{5.10}$$

The scalar functions $\underset{1}{\psi}$ and $\underset{2}{\psi}$ are two independent functions of the coordinates (Jacobian rank two). The field $F_{\nu\lambda}$ then becomes

$$F_{\nu\lambda} = 2\partial_{[\nu} \underset{1}{\psi} \partial_{\lambda]} \underset{2}{\psi} \, . \tag{5.11}$$

It should be noted that the relation between (5.10) and (5.11) is not unique. The $\underset{1}{\psi}$ and $\underset{2}{\psi}$ can be subjected to a transformation

$$\begin{aligned} \underset{1}{f} &= \underset{1}{f}(\underset{1}{\psi}, \underset{2}{\psi}) \\ \underset{2}{f} &= \underset{2}{f}(\underset{1}{\psi}, \underset{2}{\psi}) \, . \end{aligned} \tag{5.12}$$

The functions $\underset{1}{f}$ and $\underset{2}{f}$ lead to the same field $F_{\nu\lambda}$ if the Jacobian

$$\frac{\partial(\underset{1}{f}, \underset{2}{f})}{\partial(\underset{1}{\psi}, \underset{2}{\psi})} = 1 \, , \tag{5.13}$$

because

$$\partial_{[\nu}\underset{1}{f}\partial_{\lambda]}\underset{2}{f} = \frac{\partial(\underset{1}{f}, \underset{2}{f})}{\partial(\underset{1}{\psi}, \underset{2}{\psi})}\partial_{[\nu}\underset{1}{\psi}\partial_{\lambda]}\underset{2}{\psi}\,. \tag{5.14}$$

The question arises how one knows whether A_λ can be represented by the form (5.10) but not by the form (5.8). The necessary and sufficient conditions appear to be

$$\partial_{[\nu}A_{\lambda]} \neq 0\,, \tag{5.15}$$

and

$$A_{[\kappa}\partial_\nu A_{\lambda]} = 0\,. \tag{5.16}$$

Substitution of (5.10) into (5.16) indeed gives

$$\underset{1}{\psi}\partial_{[\kappa}\underset{2}{\psi}\partial_\nu\underset{1}{\psi}\partial_{\lambda]}\underset{2}{\psi} = 0\,, \tag{5.17}$$

because the function $\underset{2}{\psi}$ occurs twice in the alternation over κ, ν, and λ.

Now suppose that

$$\partial_{[\nu}A_{\lambda]} \neq 0 \tag{5.18}$$

and

$$A_{[\kappa}\partial_\nu A_{\lambda]} \neq 0\,. \tag{5.19}$$

The four-potential A_λ then cannot be represented by the form (5.8) nor by (5.10). The next simplest possibility which meets (5.18) and (5.19) is

$$A_\lambda = \underset{1}{\psi}\partial_\lambda\underset{2}{\psi} + \partial_\lambda\underset{3}{\psi}\,, \tag{5.20}$$

with $\underset{1}{\psi}$, $\underset{2}{\psi}$ and $\underset{3}{\psi}$ having a Jacobian of rank three.

Substitution of (5.20) in

$$\partial_{[\sigma}A_\kappa\partial_\nu A_{\lambda]}\,, \tag{5.21}$$

however, gives identically zero.

The representation of the four-potential by (5.20) does not yield any new physical possibilities, because fields $F_{\nu\lambda}$ resulting from (5.10) and (5.20) are the same.

The next following possibility for A_λ to be represented by gradients of scalars is

$$A_\lambda = \underset{1}{\psi}\partial_\lambda \underset{2}{\psi} + \underset{3}{\psi}\partial_\lambda \underset{4}{\psi}\,. \tag{5.22}$$

The form (5.22) is the most general one for a covariant vector in a four-dimensional manifold, because it is impossible to choose more than four independent scalars in four variables with a Jacobian of a rank higher than four. The expression (5.22) therefore corresponds to the set of relations

$$\partial_{[\nu}A_{\lambda]} \neq 0\,, \quad A_{[\kappa}\partial_\nu A_{\lambda]} \neq 0 \quad \text{and} \quad \partial_{[\sigma}A_\kappa \partial_\nu A_{\kappa]} \neq 0\,. \tag{5.23}$$

Summarizing we thus find for a field $F_{\nu\lambda} \neq 0$ that there are two physically interesting and distinct possibilities, i.e.,

$$A_\lambda = \underset{1}{\psi}\partial_\lambda \underset{2}{\psi} \tag{5.24}$$

or

$$A_\lambda = \underset{1}{\psi}\partial_\lambda \underset{2}{\psi} + \underset{3}{\psi}\partial_\lambda \underset{4}{\psi}\,. \tag{5.25}$$

Hence any arbitrary potential field can be split into two contributions

$$A_\lambda = \underset{1}{A_\lambda} + \underset{2}{A_\lambda}\,, \tag{5.26}$$

so that

$$\underset{1}{A}_{[\kappa}\partial_\nu \underset{1}{A}_{\lambda]} = \underset{2}{A}_{[\kappa}\partial_\nu \underset{2}{A}_{\lambda]} = 0\,, \tag{5.27}$$

but

$$\underset{1}{F}_{\nu\lambda} = 2\partial_{[\nu}\underset{1}{A}_{\lambda]} \neq 0 \quad \text{and} \quad \underset{2}{F}_{\nu\lambda} = 2\partial_{[\nu}\underset{2}{A}_{\lambda]} \neq 0\,, \tag{5.28}$$

with

$$2\partial_{[\nu}A_{\lambda]} = F_{\nu\lambda} = \underset{1}{F}_{\nu\lambda} + \underset{2}{F}_{\nu\lambda}\,. \tag{5.29}$$

The relation (5.29) is a decomposition of the bivector $F_{\nu\lambda}$ into two single bladed components (see SCHOUTEN [1951]).

It is of some interest to transcribe the relations (5.16) and (5.21) into a three-dimensional form

$$A_{[\nu}\partial_\nu A_{\lambda]} = 0 \begin{cases} \boldsymbol{A}\cdot\boldsymbol{B} = 0 \\ \\ \varphi\boldsymbol{B} + \boldsymbol{A} \wedge \boldsymbol{E} = 0 \end{cases} \tag{5.16a}$$

$$\partial_{[\sigma}A_\kappa\partial_\nu A_{\lambda]} = 0 \rightarrow \boldsymbol{E}\cdot\boldsymbol{B} = 0\,. \tag{5.21a}$$

The previous analysis shows that any problem which satisfies (5.21a) can be made to satisfy (5.16a) by proper choice of the potential field A_λ, because the additive component $\partial_\lambda \underset{3}{\psi}$ in (5.20) does not contribute to the field $F_{\nu\lambda}$.

By means of the permutation field (1.20) one can rewrite (5.21) in the form

$$\mathfrak{E}^{\lambda\nu\sigma\kappa}\partial_\lambda A_\nu \partial_\sigma A_\kappa\,. \tag{5.30}$$

The expression (5.30) has the appearance of a legitimate Lagrangian density. Its Euler-Lagrangian derivative, however, vanishes identically

$$\partial_\lambda \frac{\partial \mathscr{L}}{\partial(\partial_\lambda A_\nu)} = \mathfrak{E}^{\lambda\nu\sigma\kappa}\partial_\lambda\partial_\sigma A_\kappa \equiv 0\,, \tag{5.31}$$

because the components of the permutation field are constants and alternation over λ and σ causes all terms to cancel with each other.

In § 4.2 we derived the relation (4.36) from which we concluded that a structural field with an identically vanishing Lie derivative leads to an energy momentum tensor which is identically zero.

The permutation field as a structural field yields an identically vanishing Euler-Lagrange derivative as expressed by

(5.31). By applying the rules for obtaining the Lie derivative (SCHOUTEN [1951]) one may confirm that the Lie derivative of the permutation field is identically zero:

$$\underset{u}{\pounds}\mathfrak{E}^{\lambda\nu\sigma\kappa} \equiv 0\,. \tag{5.32}$$

Summarizing we can say that there are two important reasons to classify the permutation field as a trivial structural field. One is that the Euler-Lagrangian derivative of the $\mathscr{L}$ associated with the permutation field vanishes. Then secondly, the Lie derivative of the permutation field is zero which means that the energy momentum tensor associated with this $\mathscr{L}$ is identically zero.

We may refrain from investigating here to what extent these two conditions can be regarded as equivalent. Instead, we may simply state that any structural field with non-vanishing Lie derivative and leading to a nonidentically vanishing Euler-Lagrangian derivative will be called a non-trivial structural field.

It is important, however, to have a means of recognizing trivial structural fields, because it gives us an opportunity to eliminate the trivial, and therefore redundant, parts from otherwise nontrivial Lagrangians.

3. STRUCTURE OF NONTRIVIAL LAGRANGIANS

In the first section of this chapter we examined the restrictions on Lagrangians due to the fact that the relation (5.1) should be an identity in the derivatives of the Jacobian transformation elements as expressed by (5.5). It was shown that this condition resulted in the relation (4.22). One may notice that the relation (4.22) does not invoke explicitly the structural field of the Lagrangian, because we made the assumption that the structural field is a pure tensor field thus implying that its transformation only requires the elements of the Jacobian

$A^{\lambda}_{\lambda'}$ and not their derivatives. One can therefore expect another restrictive condition by requiring that (5.1) should be an identity in the elements of the Jacobian itself. To give this restrictive condition an explicit form, it is necessary to choose a specific structural field. For reasons that will become more conspicuous in chapter 9 we may choose the metrical tensor of matter-free space as the structural elements of a Lagrangian density.

Hence, for matter-free space, we assume the transformation of $\mathscr{L}$ in the form†

$$\mathscr{L}(\partial_{\nu'}A_{\lambda'}, g_{\sigma'\kappa'}) = |\Delta|^{-1}\,\mathscr{L}(\partial_{\nu}A_{\lambda}, g_{\sigma\kappa})\,, \tag{5.33}$$

because for nonconducting media we know that the potential A_{λ} itself does not occur in $\mathscr{L}$. We know already that the derivatives of the potentials occur only in the form of the curl. The transformations associated with the functional and structural fields are therefore

$$\partial_{[\nu'}A_{\lambda']} = A^{\nu\,\lambda}_{\nu'\lambda'}\partial_{[\nu}A_{\lambda]}\,, \tag{5.34}$$

$$g_{\sigma'\kappa'} = A^{\sigma\,\kappa}_{\sigma'\kappa'}g_{\sigma\kappa} \tag{5.35}$$

and in addition we know that [see (1.10)]

$$\Delta^{-1} = |A^{\sigma}_{\sigma'}| = |A^{\sigma'}_{\sigma}|^{-1}\,. \tag{5.36}$$

The condition that (5.33) should be an identity in the transformation elements $A^{\mu}_{\mu'}$ then requires that the expression

$$\frac{\partial\mathscr{L}}{\partial(\partial_{\nu'}A_{\lambda'})}\,\frac{\partial(\partial_{\nu'}A_{\lambda'})}{\partial A^{\mu}_{\mu'}} + \frac{\partial\mathscr{L}}{\partial g_{\sigma'\kappa'}}\,\frac{\partial g_{\sigma'\kappa'}}{\partial A^{\mu}_{\mu'}} = \frac{\partial\,|A^{\sigma}_{\sigma'}|}{\partial A^{\mu}_{\mu'}}\,\mathscr{L}\,, \tag{5.37}$$

should hold for any $A^{\mu}_{\mu'}$.

† In this section we write $\mathscr{L}$ instead of $\underset{0}{\mathscr{L}}$.

Using (4.24), (5.34) and (5.35), the expression (5.37) becomes

$$\mathfrak{G}^{\nu'\lambda'}(A^{\nu}_{\nu'}\delta^{\lambda}_{\mu}\delta^{\mu'}_{\lambda'} + A^{\lambda}_{\lambda'}\delta^{\nu}_{\mu}\delta^{\mu'}_{\nu'})\partial_{\nu}A_{\lambda}$$

$$+ g_{\sigma\kappa}\frac{\partial\mathscr{L}}{\partial g_{\sigma'\kappa'}}(A^{\sigma}_{\sigma'}\delta^{\kappa}_{\mu}\delta^{\mu'}_{\kappa'} + A^{\kappa}_{\kappa'}\delta^{\sigma}_{\mu}\delta^{\mu'}_{\sigma'}) = \frac{\partial\,|A^{\sigma}_{\sigma'}|}{\partial A^{\mu}_{\mu'}}\mathscr{L}. \quad (5.38)$$

The relation (5.38) is true for any $A^{\mu}_{\mu'}$: hence if $A^{\mu}_{\mu'}$ approaches the identity transformation one obtains if $\mu' \to \tau$

$$\boxed{\mathfrak{G}^{\nu\tau}2\partial_{[\nu}A_{\mu]} - \delta^{\tau}_{\mu}\mathscr{L} = -\,2g_{\nu\mu}\frac{\partial\mathscr{L}}{\partial g_{\nu\tau}}} \quad (5.39)$$

In deriving the condition (5.39), we already used the relation (4.22), which we derived earlier in the first part of this chapter. If one refrains from using (4.22) one ends up with

$$\frac{\partial\mathscr{L}}{\partial(\partial_{\nu}A_{\tau})}\partial_{\nu}A_{\mu} + \frac{\partial\mathscr{L}}{\partial(\partial_{\tau}A_{\lambda})}\partial_{\mu}A_{\lambda} - \delta^{\tau}_{\mu}\mathscr{L} = 2g_{\nu\mu}\frac{\partial\mathscr{L}}{\partial g_{\nu\tau}}, \quad (5.40)$$

which is a pure consequence of the requirement that (5.33) should be an identity in the elements $A^{\mu}_{\mu'}$. One obtains a relation of the type (5.40) if one imposes the condition of covariance with respect to the linear four-dimensional group only. It should be noted that the typical physical features of (4.22) and (5.39) are direct consequences of the conditions of general covariance. A comparison of (5.39) with (4.49) and (4.36) shows that the left-hand member of (5.39) represents the energy momentum tensor of the electromagnetic field. A further inspection of (4.36), in which the structural field was left unspecified, raises the question whether (4.36) and (5.39) are related. One can prove this to be true if one replaces the unspecified field ST by $g_{\sigma\kappa}$.

The Lie derivative of the metric is given by the expression

$$\underset{u}{\pounds} g_{\sigma\kappa} = u^\lambda \partial_\lambda g_{\sigma\kappa} + g_{\sigma\lambda}\partial_\kappa u^\lambda + g_{\lambda\kappa}\partial_\sigma u^\lambda . \tag{5.41}$$

Substituting (5.41) into (4.36) and using (4.55), one obtains

$$\partial_\nu(\mathfrak{T}_\lambda{}^\nu u^\lambda) = \frac{\partial \mathscr{L}}{\partial g_{\sigma\kappa}} u^\lambda \partial_\lambda g_{\sigma\kappa} + \frac{\partial \mathscr{L}}{\partial g_{\sigma\kappa}} (g_{\sigma\lambda}\partial_\kappa u^\lambda + g_{\lambda\kappa}\partial_\sigma u^\lambda) , \tag{5.42}$$

which means that we have equated the integrands of (4.36), because (4.36) holds for arbitrary domains of the space-time manifold. One may note that

$$\frac{\partial \mathscr{L}}{\partial g_{\sigma\kappa}} \partial_\lambda g_{\sigma\kappa} = \partial_{(\lambda)} \mathscr{L} , \tag{5.43}$$

because the explicit derivative $\partial_{(\lambda)} \mathscr{L}$ by definition [see (4.45)] operates on the structural field only.

Ordering the expression (5.42) with respect to u^λ and the derivatives of u^λ one finds

$$u^\lambda \{ \partial_\nu \mathfrak{T}_\lambda^\nu - \partial_{(\lambda)} \mathscr{L} \} + (\partial_\nu u^\lambda) \left\{ \mathfrak{T}_\lambda^\nu - 2 \frac{\partial \mathscr{L}}{\partial g_{\nu\kappa}} g_{\lambda\kappa} \right\} = 0 . \tag{5.44}$$

The expression (5.44) should be an identity in the deformation field u^λ (i.e., arbitrary u^λ and $\partial_\nu u^\lambda$). The necessary and sufficient conditions for this identity to hold require that the terms between accolades are independently zero:

$$\partial_\nu \mathfrak{T}_\lambda^\nu = \partial_{(\lambda)} \mathscr{L} \tag{5.45}$$

and

$$\mathfrak{T}_\lambda^\nu = 2 g_{\lambda\kappa} \frac{\partial \mathscr{L}}{\partial g_{\nu\kappa}} . \tag{5.46}$$

One can verify that (5.45) corresponds to (4.49), the energy momentum relation for a current and charge-free medium, and (5.46) is equivalent to the relation (5.39) which we just derived on basis of arguments of transformation compatibility.

For other structural fields than $g_{\lambda\nu}$ one can, of course, derive other relations similar to (5.39). In many important cases one can frequently guess the Lagrangian structure and then check afterwards whether the compatibility relations are satisfied. The relations (4.17) and (4.22) are always the same for A_λ as a functional field, as long as the structural field is a tensor field. The compatibility condition (5.39), however, takes different forms for different structural fields.

A decision to rule out nontensorial fields as structural elements, of course, represents a certain arbitrariness. Further investigation, in chapters 6 and following, shows that all important phenomena can be accommodated with tensorial structural fields only. It should be kept in mind that nontensorial structural fields would affect the presently accepted form of the Maxwell equations.

The investigation of so-called admissible Lagrangians is not new; much of its practical value, however, depends on the observation of some basic rules and assumptions. The important ones are the distinction between functional and structural fields and the choice of transformation behavior. The intuitive philosophy behind these rules is that any variational problem requires a reference (structure) with respect to which the variations are considered. The neglect of these rules frequently leads to a labyrinth of mathematical expressions which fail to represent meaningful physical relations.

4. SCALAR POTENTIALS

In § 2 of this chapter we have become acquainted with a set of potentials which distinguish themselves from the vector-potential and the Hertz-Fitzgerald potentials by the fact that they are true scalars for arbitrary space-time transformations. These scalars, however, have the disadvantage that the electromagnetic field itself appears as a bilinear expression in their gradients [see equations (5.11), (5.28) and (5.29)]. It is clear that the superposition principle does not hold for these scalar potentials and the question arises whether they can still satisfy linear differential equations.

The nature of these scalar potentials may be most conveniently illustrated if we restrict ourselves to cases of free space propagation directed and confined by ideally conducting surfaces. A derivation of the differential equations for the potentials requires a knowledge of the free space constitutive equations discussed in chapter 9. The free space relation between the fields is, according to (9.4),

$$\mathfrak{G}^{\lambda\nu} = \tfrac{1}{2}\underset{0}{Y} g^{\frac{1}{2}}(g^{\lambda\sigma}g^{\nu\kappa} - g^{\lambda\kappa}g^{\nu\sigma})F_{\sigma\kappa}\,. \tag{5.47}$$

The symbol $g^{\lambda\nu}$ in (5.47) is the contravariant metric tensor, g is the determinant of $g_{\lambda\nu}$ and $\underset{0}{Y}$ is a constant, the admittance of free space.

For reasons of simplicity we may consider a potential representation of $F_{\sigma\kappa}$ with two scalars only, such as given by equation (5.11). Note that the potential representation (5.11) implies that we can deal only with problems for which (5.21) holds everywhere in the field, i.e., for free space $\boldsymbol{E}\cdot\boldsymbol{B} = 0$. Substitution of (5.11) in (5.47) yields

$$\mathfrak{G}^{\lambda\nu} = \underset{0}{Y} g^{\frac{1}{2}}(g^{\lambda\sigma}g^{\nu\kappa}\partial_\sigma \underset{1}{\psi}\, \partial_\kappa \underset{2}{\psi} - g^{\lambda\kappa}g^{\nu\sigma}\partial_\sigma \underset{1}{\psi}\, \partial_\kappa \underset{2}{\psi}). \tag{5.48}$$

The Maxwell-Minkowski equations for a charge and current free medium require that the divergence of $\mathfrak{G}^{\lambda\nu}$ vanishes. The right hand member of (5.48) thus becomes [see (3.12)]

$$\underset{0}{Y}(g^{\lambda\sigma}\partial_\sigma \underset{1}{\psi}\partial_\nu g^{\frac{1}{2}}g^{\nu\kappa}\partial_\kappa \underset{2}{\psi} - g^{\lambda\kappa}\partial_\kappa \underset{2}{\psi}\partial_\nu g^{\frac{1}{2}}g^{\nu\sigma}\partial_\sigma \underset{1}{\psi})$$

$$+ \underset{0}{Y}g^{\frac{1}{2}}(g^{\nu\kappa}\partial_\kappa \underset{2}{\psi}\partial_\nu g^{\lambda\sigma}\partial_\sigma \underset{1}{\psi} - g^{\nu\sigma}\partial_\sigma \underset{1}{\psi}\partial_\nu g^{\lambda\kappa}\partial_\kappa \underset{2}{\psi}) = 0\,. \quad (5.49)$$

In the first two terms of (5.49) we may recognize the general invariant d'Alembertian (operating on a scalar) as a multiplicative factor. The structure of (5.49) becomes a little more transparent if we introduce the notations

$$g^{-\frac{1}{2}}\partial_\lambda g^{\lambda\nu}g^{\frac{1}{2}}\partial_\nu\varphi = \Box\,\varphi \quad \text{and} \quad \partial_\lambda\varphi = \varphi_\lambda, \quad g^{\lambda\nu}\partial_\nu\varphi = \varphi^\lambda\,; \qquad (5.50)$$

dropping the multiplicative factor $\underset{0}{Y}g^{\frac{1}{2}}$ the equation (5.49) then obtains the form

$$\underset{[1}{\psi^\lambda}\,\Box\,\underset{2]}{\psi} + \underset{[2}{\psi^\nu}\partial_\nu \underset{1]}{\psi^\lambda} = 0\,. \qquad (5.51)$$

Both terms in (5.51) are individually covariant for general coordinate transformations. The first term contains the familiar invariant structure of the d'Alembertian on general coordinates while the second term represents the Lie derivative of $\underset{1}{\psi^\lambda}$ with respect to $\underset{2}{\psi^\nu}$.

The behavior of (5.51) with respect to gauge transformations, however, is different. The complete form, as proven in § 2, is invariant for gauge changes of the unimodular group as defined by (5.12) and (5.13). The individual terms of (5.51), on the other hand, are not invariant for gauge transformations. This enables us to follow a similar procedure as for the vector potential, which means one imposes a gauge so that the second term vanishes. Hence the condition

$$\underset{[2}{\psi^\lambda}\partial_\lambda \underset{1]}{\psi^\nu} = 0\,, \qquad (5.52)$$

applied to the scalar potentials $\underset{1}{\psi}$ and $\underset{2}{\psi}$, may be regarded as the analogue of the Lorentz condition.

An extension of the procedure from a two scalar system ($\boldsymbol{E}\cdot\boldsymbol{B} = 0$) to a four scalar system ($\boldsymbol{E}\cdot\boldsymbol{B} \neq 0$), requires only that the second pair of potentials $\underset{3}{\psi}$ and $\underset{4}{\psi}$ satisfy a similar relation to (5.52). It is not without interest to note that a gauge change represents a transformation of the dependent variables thus generating a transformation of the gradients of the scalars without affecting the coordinates. Gauge transformations, therefore, can be considered as a special class of canonical transformations. They have the unimodular property in common.

The gauge condition (5.52) can be met automatically if the problem lends itself to an application of the principle of separation of variables. The essence of the method may be illustrated by a couple of examples which admit the use of a single bladed bivector of the field (two scalar system).

Suppose

$$\underset{1}{\psi} = \underset{1}{\psi}(t, x) \quad \text{and} \quad \underset{2}{\psi} = \underset{2}{\psi}(y, z)\,, \tag{5.53}$$

where t is the time and x, y, z the coordinates of a rectangular Cartesian frame. Substitution of (5.53) shows that (5.52) is identically satisfied for every $\underset{1}{\psi}$ and $\underset{2}{\psi}$. The equations (5.51) then reduce to the form

$$\underset{[1}{\psi^{\lambda}} \,\Box\, \underset{2]}{\psi} = 0\,, \tag{5.54}$$

or using (5.53) in (5.54) they are

$$\begin{aligned}
\underset{1}{\psi}^{0}(t, x) \;\Box\; \underset{2}{\psi}(y, z) &= 0 \\
\underset{1}{\psi}^{1}(t, x) \;\Box\; \underset{2}{\psi}(y, z) &= 0 \\
\underset{2}{\psi}^{2}(y, z) \;\Box\; \underset{1}{\psi}(t, x) &= 0 \\
\underset{2}{\psi}^{3}(y, z) \;\Box\; \underset{1}{\psi}(t, x) &= 0\,.
\end{aligned} \tag{5.54a}$$

It is clear that the system (5.54a) leads to the following differential equations for $\underset{1}{\psi}$ and $\underset{2}{\psi}$

$$\frac{1}{c^2}\frac{\partial^2}{\partial t^2}\underset{1}{\psi} - \frac{\partial^2}{\partial x^2}\underset{1}{\psi} = 0 \tag{5.55}$$

$$\frac{\partial^2}{\partial x^2}\underset{2}{\psi} + \frac{\partial^2}{\partial y^2}\underset{2}{\psi} = 0\,. \tag{5.56}$$

The boundary conditions are determined from the expressions for the field vectors which follow from (5.11)

$$\begin{aligned} E_1 &\sim \partial_0 \underset{[1}{\psi} \partial_1 \underset{2]}{\psi} = 0 & H_1 &\sim \partial_2 \underset{[1}{\psi} \partial_3 \underset{2]}{\psi} = 0 \\ E_2 &\sim \partial_0 \underset{[1}{\psi} \partial_2 \underset{2]}{\psi} \neq 0 & H_2 &\sim \partial_3 \underset{[1}{\psi} \partial_1 \underset{2]}{\psi} \neq 0 \\ E_3 &\sim \partial_0 \underset{[1}{\psi} \partial_3 \underset{2]}{\psi} \neq 0 & H_3 &\sim \partial_1 \underset{[1}{\psi} \partial_2 \underset{2]}{\psi} \neq 0\,. \end{aligned} \tag{5.57}$$

The field expressions (5.57) show that the field vectors are always transverse with respect to the x direction, thus implying that the assumption (5.53) lends itself to the description of an ideal transverse electromagnetic wave (TEM wave) in the x direction, for instance a wave in a transmission line. Note that the two-scalar assumption (5.53) is not adequate for a lossy line, because a lossy line has an electric field component in the direction of propagation and is, therefore, not an ideal TEM wave. The reader may find that the solution of (5.55) and (5.56) for a given geometry, a coaxial line say, can be completed by standard procedures in conjunction with the boundary conditions following from (5.57).

A still simpler example, which applies to a transverse electric wave propagating between two parallel, ideally conducting plates, is

$$\underset{1}{\psi} = Az \qquad A \text{ is a constant}$$
$$\underset{2}{\psi} = e^{i(\omega t - kx)}\, Y(y)\,. \tag{5.58}$$

The scalar $\underset{1}{\psi}$ is of course a trivial solution of the d'Alembertian. Following the same procedure one finds from $\Box \underset{2}{\psi} = 0$,

$$\begin{cases} Y''(y) + \gamma^2 Y(y) = 0 \\ \text{with } \gamma^2 = \dfrac{\omega^2}{c^2} - k^2\,. \end{cases} \tag{5.59}$$

The fields are

$$\begin{aligned} E_1 &= 0\,, & H_1 &\sim AY'(y)\, e^{i(\omega t - kx)} \\ E_2 &= 0\,, & H_2 &\sim ikAY(y)\, e^{i(\omega t - kx)} \\ E_3 &\sim i\omega AY(y)\, e^{i(\omega t - kx)}\,, & H_3 &= 0\,. \end{aligned} \tag{5.60}$$

Suppose that the conducting planes are parallel to the x, z plane with a mutual separation a, the boundary conditions then become

$$E_3 = 0 \text{ for } y = 0 \text{ and } y = a \quad \text{or} \quad Y(0) = Y(a) = 0\,. \tag{5.61}$$

The solutions of (5.59) satisfying (5.61) are

$$Y(y) = \sin \gamma y, \quad \text{with } \gamma = n\pi/a, \quad n = 1, 2 \ldots$$

Hence

$$\boxed{\frac{n^2\pi^2}{a^2} = \frac{\omega^2}{c^2} - k^2} \tag{5.62}$$

The frequency relation (5.62) describes the cut-off and dispersion properties of an electromagnetic duct. If we define $\omega/c = k_0$ as the free space wave number, $\omega/k = u$ as the phase velocity in the duct and $d\omega/dk = g$ as the group velocity, we obtain the following familiar expressions for the propagation behavior in the duct

$$\frac{k}{k_0} = \sqrt{1 - \frac{n^2\pi^2}{a^2k_0^2}},$$

$$u = \frac{c}{\sqrt{1 - \frac{n^2\pi^2}{a^2k_0^2}}}, \qquad g = c\sqrt{1 - \frac{n^2\pi^2}{a^2k_0^2}}.$$

CHAPTER 6

GENERAL PROPERTIES OF THE MEDIUM

1. GENERAL

Thus far, we have primarily discussed the field equations and general field relations invoked by the Lagrangian density. The Lagrangian, however, was not specified in any detail, except that it had to obey certain restrictions that were associated with the general feasibility of the generally covariant variational formulation. Those restrictions do not represent physical properties of the medium. Specific properties of the electromagnetic medium, which one expects to be determined by the explicit Lagrangian or the constitutive equations, have been discussed only for interpretative purposes. For general orientation we may mention some of the most conspicuous properties of electromagnetic media and their associated phenomena. We shall restrict this tabulation to nonconducting media, which means that we will consider in particular the phenomenological aspects of dielectric and magnetic properties of the medium.

Within the realms of nonconducting matter one may distinguish between the following variety of characteristics. They are listed below pairwise so as to contrast the simple with the more complex features:

I	linear	A	nonlinear
II	nondissipative	B	dissipative
III	isotropic	C	anisotropic
IV	reciprocal	D	nonreciprocal
V	uniform	E	nonuniform
VI	nondispersive	F	dispersive

The combination (I, II, III, IV, V, VI) represents the simplest possible medium (e.g., ideal free space). The combination (A, B, C, D, E, F) represents the most complex case. In the following chapters we propose to discuss some intermediate cases, e.g.,

(I, II, C, IV, E, VI)	chapter 7,
(I, II, C, D, V, F)	chapter 8,
(I, II, III, D, E, VI)	chapter 9.

We may now proceed discussing point-wise typical physical phenomena associated with the properties mentioned under the capital symbols.

A. Nonlinearity causes frequency conversion and may ultimately lead to energy degradation.

B. Dissipative behavior is characterized by a transition from ordered into disordered energy. Nonlinearity can be a mechanism which brings about dissipation. The mutual relation of A and B will not be investigated here.

C. Anisotropy leads to two physical phenomena known as birefringence and natural optical activity.

The birefringence depends on the direction of propagation and the rotational symmetry of the medium. Optical activity can occur in media with complete rotational symmetry, but without a centre of symmetry. We will find later that optical activity is an essentially dispersive phenomenon.

D. Nonreciprocity in nonconductive media is represented by three effects. They are the Fresnel-Fizeau effect in moving media and the two Faraday effects, i.e., the dielectric and the magnetic Faraday effect. The latter two are essentially dispersive effects, the Fresnel-Fizeau effect is not. The two Faraday effects may occur simultaneously. Roughly speaking one can say that the effect originally investigated by Faraday is of the dielectric type (BORN [1933]). The Faraday effect in ferrites, as more recently applied in microwave structures, is primarily of the magnetic type (POLDER [1949]).

It should be noted that each of the nonreciprocal bulk effects in nonconductive media is generated by biasing fields: a velocity field for the Fresnel-Fizeau effect and a magnetic field for the Faraday effects. They can be regarded as induced anisotropies in a space-time sense, because the biasing fields remove the time symmetry of the medium.

E. Nonuniformity can be associated with the phenomenon of scattering due to the presence of centres with different dielectric (and magnetic) properties causing reflection and diffraction of waves. It is important to distinguish between nonuniformity of a random nature and periodical nonuniformity as occurring in lattices.

The definition of uniformity invokes a subtle little point which is usually not explicitly mentioned, because of *a priori* and tacit assumptions about the uniformity of the underlying space. Any criterion for uniformity requires a suitable reference. Assuming the ideal free space as the reference, one may identify its uniformity with its Euclidian nature. Then, knowing that Euclidian space admits Cartesian (or uniform) coordinates throughout finite domains, one can give the following unambiguous definition of a uniform medium:

A medium is called uniform if the explicit derivatives of the Lagrangian density with respect to Cartesian coordinates vanish.

An equivalent definition is: the components of the structural fields of a uniform medium are constants if expressed with respect to Cartesian coordinates.

The following point should be noted in order to prevent any confusion when comparing a generally covariant formulation with a formulation which is covariant with respect to the group of rotations only: the generally covariant form of the constitutive equations itself cannot tell us whether a medium is uniform or anisotropic unless one specifies the nature of the underlying space and its coordinates.

This statement brings out a partial duality between nonuniformity and anisotropy on the one hand and the geometrical nature of space and its coordinates on the other hand. Even if, for all practical purposes, space can be assumed to be flat, then general covariance is still a mathematical physical necessity for a proper non-ad hoc formulation of problems of nonuniformity, anisotropy and general coordinates.

F. As mentioned earlier, dispersion is associated with noninstantaneous and nonlocal relations between the (macroscopic) electromagnetic fields. In more physical terms, the polarization response of a medium, as a rule, is not instantaneous due to the inertia of the microphysical charges in the medium. In addition, one expects that the polarization at a particular point in a medium may partly be determined by the dipole fields in the neighborhood of that point.

The traditional real algebraic relation between the fields is not adequate to represent dispersion even if one makes the coefficients ε and μ functions of the frequency or wave number, because the phase shift between cause and effect is not accounted for by a real algebraic relation. It was noticed quite early, as in circuit theory, that the formalism of complex field variables enables one to remove this inadequacy.

Complex field relations can be regarded as (real) integral relations in disguise. The integrals sum the influences originating from different space-time points in the medium, thus telling the more precise story of noninstantaneous and nonlocal interaction.

An example of such an integral relation representing a noninstantaneous, linear interaction between E and D is, for instance,

$$D(t) = \int_{-\infty}^{t} \varepsilon(t, \tau)\, E(\tau)\, \mathrm{d}\tau\,. \tag{6.1}$$

If one cares to include nonlocal interaction (for simplicity in one dimension) one may consider the form

$$D(t, x) = \int_{-\infty}^{t} \mathrm{d}\tau \int_{x-c(t-\tau)}^{x+c(t-\tau)} \mathrm{d}\xi \; \varepsilon(t, \tau; \xi, x)\, E(\tau, \xi)\,. \qquad (6.2)$$

The limits of integration in (6.1) and (6.2) have been chosen so as to meet the condition of causality, assuming the free-space velocity c as the highest possible velocity of interaction between different points of the medium.

It is possible and desirable to consider causal behavior as a property of the medium. One may expect, for instance, that a passive medium should behave in a causal manner, so that effect $D(t)$ cannot precede the cause $E(t)$. Conversely, it is not true that a nonpassive medium is necessarily noncausal. Vacuum tube amplifiers are examples of nonpassive discrete elements exhibiting causal behavior, provided they are stable. These notions for a lumped network can be reinterpreted in terms of field properties of a nonpassive medium. The interaction space of microwave amplifier devices may be regarded as an example of an active medium that can be stable.

The kernel $\varepsilon(t, \tau)$ in (6.1) represents only the noninstantaneous interaction properties of the medium. Expressing causality as a property of $\varepsilon(t, \tau)$, one imposes the condition

$$\varepsilon(t, \tau) = 0 \quad \text{for} \quad \tau > t\,. \qquad (6.3)$$

The expression (6.1) now can be given a form with the upper limit of integration equal to infinity

$$D(t) = \int_{-\infty}^{\infty} \varepsilon(t, \tau)\, E(\tau)\, \mathrm{d}\tau\,. \qquad (6.4)$$

One can stipulate another practical detail in the structure of the kernel if the response properties themselves do not depend on the time. For such a medium one imposes the condition of translational invariance in the time i.e.,

$$D(t + t_0) = \int_{-\infty}^{\infty} \varepsilon(t, \tau)\, E(\tau + t_0)\, d\tau\,, \tag{6.5}$$

for arbitrary t and t_0.

A straightforward change of variables shows that (6.5) holds if the kernel has the form

$$\varepsilon(t, \tau) = \varepsilon(t - \tau)\,. \tag{6.6}$$

Hence summarizing:

The polarization properties of a linear, isotropic, stationary medium with a local but noninstantaneous response, independent of the time itself, can now be represented by the integral relation

$$D(t) = \int_{-\infty}^{\infty} \varepsilon(t - \tau)\, E(\tau)\, d\tau\,. \tag{6.7}$$

The condition of causality (6.3) requires that

$$\varepsilon(t - \tau) = 0 \quad \text{for} \quad \tau > t\,. \tag{6.8}$$

Taking the Fourier transform of (6.7) one can transcribe this integral relation by means of the Borel theorem into an equivalent (complex) algebraic form for the frequency (ω) domain

$$D(\omega) = \varepsilon(\omega)\, E(\omega)\,, \tag{6.7a}$$

in which the complex quantities $D(\omega)$, $\varepsilon(\omega)$ and $E(\omega)$ are the (complex) Fourier transforms of the real quantities $D(t)$, $\varepsilon(t)$ and $E(t)$.

It should be mentioned that the function $\varepsilon(t)$ can have rather unconventional analytical properties. For the limiting case of a direct instantaneous response, it becomes a Dirac "delta" function. The customary Fourier arguments then break down. Nevertheless, one can show that there is an important class of functions $\varepsilon(t)$ such that the Fourier integral

$$\varepsilon(\omega) = \int_{-\infty}^{\infty} \varepsilon(t)\, \mathrm{e}^{\mathrm{i}\omega t}\, \mathrm{d}t$$

represents a function in ω which has an analytic continuation in the upper half of the complex frequency plane ω. The function $\varepsilon(\omega)$ can then be represented by a Cauchy contour integral around the upper half of the frequency plane

$$\int_{c} \frac{\varepsilon(s)}{s-\omega}\, \mathrm{d}s\,.$$

The integral over the infinite semicircle of the contour vanishes, because of the exponential factor $\mathrm{e}^{\mathrm{i}\omega t}$ in the integral representation of $\varepsilon(\omega)$. The single pole $s = \omega$ occurs on the real axis and should therefore be counted with half a residue. Equating the real and imaginary parts of the contour integral with those of half the residue of $\varepsilon(s)$, i.e., $\mathrm{i}\pi\varepsilon(\omega)$, one finds that the real (Re) and imaginary (Im) parts of $\varepsilon(\omega)$ are related by the so-called Hilbert transforms

$$\mathrm{Re}\ \varepsilon(\omega) = \frac{1}{\pi} \mathrm{P} \int_{-\infty}^{\infty} \frac{\mathrm{Im}\ \varepsilon(s)}{s-\omega}\, \mathrm{d}s$$

$$\mathrm{Im}\ \varepsilon(\omega) = -\frac{1}{\pi} \mathrm{P} \int_{-\infty}^{\infty} \frac{\mathrm{Re}\ \varepsilon(s)}{s-\omega}\, \mathrm{d}s\,. \tag{6.8a}$$

The integrals (6.8a) are principle value (P) integrals, implying that the discontinuity of the integrand should be approached symmetrically from either side. The integrals (6.8a) can be converted into an equivalent set, known as the Kramers-Kronig relations, if one uses the even and odd properties of the real and imaginary parts of $\varepsilon(\omega)$ [following from the reality of $\varepsilon(t)$], i.e.,

$$\mathrm{Re}\,\varepsilon(\omega) = +\frac{2}{\pi}\,\mathrm{P}\int_0^\infty \frac{s\,\mathrm{Im}\,\varepsilon(s)}{s^2-\omega^2}\,\mathrm{d}s$$

$$\mathrm{Im}\,\varepsilon(\omega) = -\frac{2\omega}{\pi}\,\mathrm{P}\int_0^\infty \frac{\mathrm{Re}\,\varepsilon(s)}{s^2-\omega^2}\,\mathrm{d}s\,. \tag{6.8b}$$

The integrals (6.8a) and (6.8b), in a sense, but with considerable reservation, can be regarded as a transcription of the causality requirement (6.8). One can express oneself in a slightly more precise way by saying that (6.8a) and (6.8b) define an important class of causal response functions $\varepsilon(\omega)$. Response functions $\varepsilon(\omega)$ which do not satisfy (6.8a) or (6.8b) are not necessarily noncausal. In this connection we should recall that the relations (6.7), (6.8), etc., are based on an assumption that the medium shows noninstantaneous but local interaction of the fields. Dispersion associated with nonlocal behavior [see (6.2)] obeys other laws which are different depending whether the medium has a randomly (ROSENFELD [1951]) or a periodically (BRILLOUIN [1947]) discrete structure.

The discussion thus far covers the case of an isotropic medium. It is easy to remove this restriction. Instead of (6.7a) one obtains a matrix equation

$$D_l(\omega) = \sum_k \varepsilon_{lk}(\omega)\,E_k(\omega)\,, \tag{6.9}$$

with the conclusion that each of the elements $\varepsilon_{lk}(\omega)$ individually should obey the Kramers-Kronig relations (6.8b).

The nondissipative behavior of the medium is expressed by the Hermitian symmetry of the matrix ε_{lk},

$$\overset{*}{\varepsilon}_{lk} = \varepsilon_{kl} , \tag{6.10}$$

in which the * denotes the complex conjugate.

The analytical treatment of causality and dispersion is full of pitfalls and unexpected dangers. The reader may do well to complete this sketchy presentation with other more complete sources of information. A textbook treatment can be found in LANDAU-LIFSCHITZ [1957] and HILGEVOORD [1960]. The basic mathematical theorems on causality for L_2 functions (i.e., excluding improper functions) are given by TITCHMARSH [1948] (theorems 93 and 95). Relevant information from the realms of circuit theory can be found in the books by BODE [1945] and GUILLEMIN [1949].

2. THE COVARIANT FORM OF THE LINEAR CONSTITUTIVE EQUATIONS

The generally covariant analogue of the customary constitutive relations

$$\begin{aligned} \boldsymbol{B} &= \underset{\mathrm{r}}{\mu}\underset{0}{\mu}\boldsymbol{H} \\ \boldsymbol{D} &= \underset{\mathrm{r}}{\varepsilon}\underset{0}{\varepsilon}\boldsymbol{E} , \end{aligned} \tag{6.11}$$

can be written in the form

$$\mathfrak{G}^{\lambda\nu} = \tfrac{1}{2}\chi^{\lambda\nu\sigma\kappa}F_{\sigma\kappa} . \tag{6.12}$$

The coefficients $\chi^{\lambda\nu\sigma\kappa}$ constitute a tensor density of weight $+1$ which can be used to describe isotropic as well as anisotropic matter. The factor $\frac{1}{2}$ has been added to compensate for the double summation over σ and κ so as to ascertain a maximum of conformity with the conventional form (6.11).

The number of components of the constitutive tensor is restricted by the skew symmetry in the two index pairs λ,ν and σ,κ

$$\begin{aligned} \chi^{\lambda\nu\sigma\kappa} &= -\chi^{\lambda\nu\kappa\sigma} \\ \chi^{\lambda\nu\sigma\kappa} &= -\chi^{\nu\lambda\sigma\kappa} \end{aligned} \tag{6.13}$$

following from the skew symmetry of the fields $\mathfrak{G}^{\lambda\nu}$ and $F_{\sigma\kappa}$.

In addition to the relations (6.13), one can expect the symmetry

$$\chi^{\lambda\nu\sigma\kappa} = \chi^{\sigma\kappa\lambda\nu}. \tag{6.14}$$

The relation (6.14) follows from the definition equation (4.24) of the field $\mathfrak{G}^{\lambda\nu}$,

$$\mathfrak{G}^{\lambda\nu} = \frac{\partial \mathscr{L}}{\partial(\partial_\lambda A_\nu)} = 2\frac{\partial \mathscr{L}}{\partial F_{\lambda\nu}}, \tag{6.15}$$

if one considers that

$$\frac{\partial \mathscr{L}}{\partial F_{\lambda\nu} \partial F_{\sigma\kappa}} = \frac{\partial \mathscr{L}}{\partial F_{\sigma\kappa} \partial F_{\lambda\nu}}. \tag{6.16}$$

The Lagrangian corresponding to the constitutive equation (6.12) then is

$$\mathscr{L} = \tfrac{1}{8}\chi^{\lambda\nu\sigma\kappa} F_{\lambda\nu} F_{\sigma\kappa}. \tag{6.17}$$

It follows from the symmetry conditions (6.13) and (6.14) that the constitutive tensor can be represented by a six-by-six symmetric matrix (21 elements).

A further important restriction can be added, which reduces the number of independent elements to 20. Let us consider, for instance, the case of a uniform medium on a Cartesian frame. The alternating components in the constitutive tensor then lead to an identically vanishing contribution in the Euler-Lagrangian derivative, as discussed in chapter 5 [see (5.31)]. Hence, in a uniform medium on a Cartesian Lorentz frame, one can impose the condition

$$\chi^{[\lambda\nu\sigma\kappa]} = 0 \,. \tag{6.18}$$

The equation (6.18), however, has the generally invariant form and one is tempted to inquire what happens if one removes the restriction of uniformity. The coefficients $\chi^{\lambda\nu\sigma\kappa}$ are then functions of the coordinates. Performing the operation with the Euler-Lagrangian derivative, one finds that the contribution of the alternating part still vanishes for a non-uniform medium if

$$\partial_\nu \chi^{[\lambda\nu\sigma\kappa]} = 0 \,. \tag{6.19}$$

The equation (6.19), like (6.18), again has the generally invariant form, because the divergence of any alternating contravariant density of weight $+1$ is known to be a natural invariant (§ 3.2).

An inspection of (6.18) and (6.19) shows that the only coefficients occurring in the alternation are associated with the relations between the magnetic and electric fields, i.e., between $\boldsymbol{D}$ and $\boldsymbol{B}$ and between $\boldsymbol{H}$ and $\boldsymbol{E}$. Coefficients associated with permeability and permittivity do not occur in (6.18) nor in (6.19).

It is known, as will be discussed more extensively later, that physically a local instantaneous coupling between electric and magnetic fields in stationary matter has never been found. Therefore, the relevant coefficients of $\chi^{\lambda\nu\sigma\kappa}$ vanish in a frame of reference fixed in the medium. Hence, if (6.18) and (6.19) hold for stationary matter, then the general covariance guarantees their general validity for any frame of reference.

Consequently, (6.18) and (6.19) are generally valid without the restriction of uniformity.

Epitomizing the present conclusions, we have that the linear and real constitutive equation

$$\mathfrak{G}^{\lambda\nu} = \tfrac{1}{2}\chi^{\lambda\nu\sigma\kappa}F_{\sigma\kappa} \tag{6.12}$$

describes in a generally covariant manner the electromagnetic behavior of a general linear medium with instantaneous and local interaction between the fields.

The constitutive tensor $\chi^{\lambda\nu\sigma\kappa}$ obeys the following symmetry relations (which are independent of the symmetry properties of the medium):

$$\left\{\begin{aligned} \chi^{\lambda\nu\sigma\kappa} &= -\chi^{\nu\lambda\sigma\kappa}, \\ \chi^{\lambda\nu\sigma\kappa} &= -\chi^{\lambda\nu\kappa\sigma}, \end{aligned}\right. \tag{6.13}$$

$$\chi^{\lambda\nu\sigma\kappa} = \chi^{\sigma\kappa\lambda\nu}, \tag{6.14}$$

$$\chi^{[\lambda\nu\sigma\kappa]} = 0\,. \tag{6.18}$$

These relations reduce the number of independent elements of the constitutive tensor to 20.

The constitutive tensor in addition satisfies the covariant differential expression

$$\partial_\nu\chi^{[\lambda\nu\sigma\kappa]} = 0\,. \tag{6.19}$$

The constitutive tensor transforms as a contravariant tensor density of weight $+1$, according to the law

$$\chi^{\lambda'\nu'\sigma'\kappa'} = |\Delta|^{-1}A^{\lambda'\nu'\sigma'\kappa'}_{\lambda\;\nu\;\sigma\;\kappa}\,\chi^{\lambda\nu\sigma\kappa}. \tag{6.20}$$

It is possible to introduce complex elements in the equation

(6.12) in order to extend the conclusions to dispersive media in the sense as discussed in the previous section. The elements of the constitutive tensor are then functions of the components of the frequency wave vector. For stationary matter and local interaction they reduce to functions of the frequency only. The nondissipative behavior of the medium is expressed by the Hermitian symmetry† of the matrix associated with the constitutive tensor, which means that (6.14) should be modified, i.e.,

$$\mathrm{Re}\,\chi^{\lambda\nu\sigma\kappa} = \mathrm{Re}\,\chi^{\sigma\kappa\lambda\nu} \tag{6.14a}$$

$$\mathrm{Im}\,\chi^{\lambda\nu\sigma\kappa} = -\,\mathrm{Im}\,\chi^{\sigma\kappa\lambda\nu}. \tag{6.14b}$$

All the other symmetry relations remain unchanged, including (6.18).

The statement that a coupling between electric and magnetic fields is absent in a stationary medium is not true any more for nonlocal and noninstantaneous interaction. We will see later that media with natural optical activity do have such coupling as can be expected from the helical structure of their molecules. The coefficients characterizing the optical activity are purely imaginary and therefore subject to the relation (6.14b). The reader is invited to check for the imaginary coefficients that (6.14b) complies with (6.18).

Using the identification table 2 (row *C*) in chapter 3, one obtains the following matrix form for the elements of the constitutive tensor:

† The proof of the Hermitian symmetry of this six-by-six matrix follows from the expression (4.57) for $d\mathscr{L}$, which differs by a total differential from the differential of the field energy density, and from the condition that the time average of $\mathscr{L}$ over one period should vanish for an absorption-free medium. Compare the analogous proof for the dielectric tensor in LANDAU and LIFSCHITZ [1957] § 76 and § 82.

$\chi^{\lambda\nu\sigma\kappa}$	01	02	03	23	31	12
	$-E_1$	$-E_2$	$-E_3$	$\tilde{B}_1$	$\tilde{B}_2$	$\tilde{B}_3$
01 D_1	$-\varepsilon_{11}$	$-\varepsilon_{12}$	$-\varepsilon_{13}$	$\tilde{\gamma}_{11}$	$\tilde{\gamma}_{12}$	$\tilde{\gamma}_{13}$
02 D_2	$-\overset{*}{\varepsilon}_{12}$	$-\varepsilon_{22}$	$-\varepsilon_{23}$	$\tilde{\gamma}_{21}$	$\tilde{\gamma}_{22}$	$\tilde{\gamma}_{23}$
03 D_3	$-\overset{*}{\varepsilon}_{13}$	$-\overset{*}{\varepsilon}_{23}$	$-\varepsilon_{33}$	$\tilde{\gamma}_{31}$	$\tilde{\gamma}_{32}$	$\tilde{\gamma}_{33}$
23 $\tilde{H}_1$	$\overset{*}{\tilde{\gamma}}_{11}$	$\overset{*}{\tilde{\gamma}}_{21}$	$\overset{*}{\tilde{\gamma}}_{31}$	χ_{11}	χ_{12}	χ_{13}
31 $\tilde{H}_2$	$\overset{*}{\tilde{\gamma}}_{12}$	$\overset{*}{\tilde{\gamma}}_{22}$	$\overset{*}{\tilde{\gamma}}_{32}$	$\overset{*}{\chi}_{12}$	χ_{22}	χ_{23}
12 $\tilde{H}_3$	$\overset{*}{\tilde{\gamma}}_{13}$	$\overset{*}{\tilde{\gamma}}_{23}$	$\overset{*}{\tilde{\gamma}}_{33}$	$\overset{*}{\chi}_{13}$	$\overset{*}{\chi}_{23}$	χ_{33}

(6.21)

This matrix form (6.21) accounts for the symmetry conditions (6.13), (6.14a) and (6.14b). The requirement (6.18) expressed in the elements of the matrix (6.21) is

$$\tilde{\gamma}_{11} + \tilde{\gamma}_{22} + \tilde{\gamma}_{33} + \overset{*}{\tilde{\gamma}}_{11} + \overset{*}{\tilde{\gamma}}_{22} + \overset{*}{\tilde{\gamma}}_{33} = 0\,. \qquad (6.18b)$$

It is seen that the six-by-six matrix (6.21) splits into four three-by-three submatrices. The diagonal submatrices represent the permittivity ε_{lk} and the inverse of the permeability χ_{lk}. Their real coefficients are associated with birefringence, the imaginary ones with the Faraday rotation.

The real parts of the matrices $\tilde{\gamma}_{lk}$ and $\tilde{\gamma}^*_{kl}$ represent physically the Fresnel-Fizeau effect, which states that light waves participate in the motion of the medium in the ratio $(1 - 1/n^2)$; n being the refractive index. It is possible to illustrate this by subjecting the constitutive tensor for an isotropic stationary medium,

$\chi^{\lambda\nu\sigma\kappa}$	$-\boldsymbol{E}$	$\boldsymbol{B}$
$\boldsymbol{D}$	$-\varepsilon_r\varepsilon_0$	0
$\boldsymbol{H}$	0	$(\mu_r\mu_0)^{-1}$

(6.22)

to the infinitesimal Lorentz transformation (3.59). The result of the transformation (6.20) on (6.22) is then

$\chi^{\lambda'\nu'\sigma'\kappa'}$		$-\boldsymbol{E}$ 01	02	03	$\boldsymbol{B}$ 23	31	12
	01	$-\varepsilon_r\varepsilon_0$	0	0	0	$-\varepsilon_r\varepsilon_0\alpha v_3$	$\varepsilon_r\varepsilon_0\alpha v_2$
$\boldsymbol{D}$	02	0	$-\varepsilon_r\varepsilon_0$	0	$\varepsilon_r\varepsilon_0\alpha v_3$	0	$-\varepsilon_r\varepsilon_0\alpha v_1$
	03	0	0	$-\varepsilon_r\varepsilon_0$	$-\varepsilon_r\varepsilon_0\alpha v_2$	$\varepsilon_r\varepsilon_0\alpha v_1$	0
	23	0	$\varepsilon_r\varepsilon_0\alpha v_3$	$-\varepsilon_r\varepsilon_0\alpha v_2$	$(\mu_r\mu_0)^{-1}$	0	0
$\boldsymbol{H}$	31	$-\varepsilon_r\varepsilon_0\alpha v_3$	0	$\varepsilon_r\varepsilon_0\alpha v_1$	0	$(\mu_r\mu_0)^{-1}$	0
	12	$\varepsilon_r\varepsilon_0\alpha v_2$	$-\varepsilon_r\varepsilon_0\alpha v_1$	0	0	0	$(\mu_r\mu_0)^{-1}$

(6.23)

with $\alpha = (1 - 1/\varepsilon_r\mu_r)$ and ε_r and μ_r the relative permittivity and relative permeability. The matrix (6.23) shows that the pseudo tensor $\tilde{\gamma}_{lk}$ is antisymmetric, corresponding to the fact that the translating medium has rotational symmetry around the direction of motion (v_1, v_2, v_3). The transformed matrix (6.23) is invariant with respect to an infinitesimal Lorentz transformation for $\alpha = 0$; which occurs, e.g., if $\underset{r}{\varepsilon} = \underset{r}{\mu} = 1$. The same, of course, is true for general Lorentz transformations. The evaluation of the transformations (6.20), however, is then much more cumbersome.

The imaginary elements in $\tilde{\gamma}_{lk}$ and $\tilde{\gamma}^*_{lk}$ will be shown to represent the natural optical activity. In this connection it is to be noted that $\tilde{\gamma}_{lk}$ and $\tilde{\gamma}^*_{lk}$, as three-dimensional pseudo tensors, are not invariant for a spatial inversion, because they connect a polar vector with an axial vector [see § 2.6, equation (2.52)]. The $\tilde{\gamma}_{lk}$, therefore, can only occur in a medium without a centre of symmetry. The permittivity ε_{lk} and the inverse permeability, as to be expected, are not affected by a spatial inversion, because they connect either two polar vectors or two axial vectors.

3. THE COVARIANT FORM OF THE WAVE EQUATION

The constitutive equation (6.12)

$$\mathfrak{G}^{\lambda\nu} = \tfrac{1}{2}\chi^{\lambda\nu\sigma\kappa}F_{\sigma\kappa}\,, \tag{6.24}$$

as discussed in the previous section, is in a general covariant form, if one adheres to the transformation rule (6.20). This equation can be expressed as a relation between the field $\mathfrak{G}^{\lambda\nu}$ and the derivatives of the four-potential by means of (3.36), i.e.,

$$F_{\sigma\kappa} = 2\partial_{[\sigma}A_{\kappa]}\,. \tag{6.25}$$

Substitution of (6.25) into (6.24) yields

$$\mathfrak{G}^{\lambda\nu} = \chi^{\lambda\nu\sigma\kappa}\partial_\sigma A_\kappa\,. \tag{6.26}$$

The alternation sign has been dropped because of the summation over the indices σ and κ, which happen to be antisymmetric according to (6.13). The equation (6.26) still has the generally invariant form, because (6.25) is a natural differential invariant.

The field $\mathfrak{G}^{\lambda\nu}$ is known to satisfy the Minkowski-Maxwell equation (3.12), i.e.,

$$\partial_\nu \mathfrak{G}^{\lambda\nu} = \mathfrak{c}^\lambda . \tag{6.27}$$

Hence substitution of (6.26) into (6.27) yields an equation of the form

$$\boxed{\partial_\nu \chi^{\lambda\nu\sigma\kappa} \partial_\sigma A_\kappa = \mathfrak{c}^\lambda} \tag{6.28}$$

The equation (6.28) really represents a set of four simultaneous partial differential equations of the second order in the components of the four-potential A_κ. Its form invariance is determined by the invariance of (6.27) and (6.26), which was extensively discussed in chapter 3. It should be noted that the invariance of the equation (6.28) does not depend on the use of covariant derivatives.

The equation (6.28) can be regarded as a generalized covariant form of the conventional wave equations of the type (3.2) and (3.5). It should be kept in mind that it is not possible to make a transition from (3.2) to (6.28) by means of the recipe of covariant derivatives. The equation (6.28) is admittedly more complex, but it does cover a much wider field of application than the isotropic versions (3.2) and (3.5).

The usual derivation of the customary wave equation in vacuo of the type (3.2) for the potentials invokes an assumption that is known as the Lorentz condition [see (3.38)]

$$\operatorname{div} \boldsymbol{A} + \underset{0}{\varepsilon}\underset{0}{\mu} \frac{\partial \varphi}{\partial t} = 0 . \tag{6.29}$$

The generally covariant analogue of this equation can be

written in the form

$$g^{\lambda\nu}\nabla_\nu A_\lambda = 0 . \tag{6.30}$$

Please note that it is not possible to write (6.30) in the form of a natural invariant, without the use of the metric. An equivalent form that has the outward appearance of a natural invariant can be obtained by introducing a potential density vector

$$\mathfrak{A}^\kappa = | g |^{\frac{1}{2}} g^{\kappa\lambda} A_\lambda , \tag{6.31}$$

with g the determinant of the metric tensor $g_{\lambda\nu}$. The equation (6.30) then takes the form

$$\partial_\kappa \mathfrak{A}^\kappa = 0 . \tag{6.32}$$

The present considerations, however, are applicable only for the vacuum. The Lorentz condition can be extended in the conventional form to an isotropic medium at rest. The ε_0 and the μ_0 in (6.29) then should be replaced by the ε and μ of the medium. It is obvious that the generalizations (6.30) and (6.32) then should be submitted to a corresponding modification which implies the introduction of a tensor $g_{\lambda\nu}$ distinct from the usual metric tensor. Generalizations of this kind for an anisotropic and moving medium, however, do not seem to be feasible; therefore we abstain from a general use of the Lorentz condition.

Imposing a condition of the kind (6.29), (6.30) or (6.32) is a mathematical step to reduce the physical redundance inherent in the potentials (see, e.g., § 5.2). An inspection of the wave equation (6.28) shows why it is useful to impose the condition (6.30). A plane wave solution

$$A_\lambda = \hat{A}_\lambda \, e^{ik_\nu x^\nu} , \tag{6.33}$$

reduces the condition (6.30) to the form

$$g^{\lambda\nu}\hat{A}_{\lambda}k_{\nu} = 0\,, \tag{6.34}$$

which means that the potential vector can be taken perpendicular to the frequency-wave vector in a four-dimensional sense. Substitution of (6.33) in (6.28) (with $\mathfrak{c}^{\lambda} = 0$) yields as a first order result, assuming approximate homogeneity,

$$\chi^{\lambda\nu\sigma\kappa}k_{\nu}k_{\sigma}\hat{A}_{\kappa} = 0\,. \tag{6.35}$$

It is clear that (6.35) is identically satisfied for any $\hat{A}_{\kappa}$ or component in $\hat{A}_{\kappa}$ which is parallel to k_{σ} because alternation over σ and κ gives zero. Therefore, one may impose a condition that $\hat{A}_{\kappa}$ has no components in the direction of k_{σ}. These results are exact only for a uniform medium on a linear frame of reference.

The preceding considerations provide some insight for a proper evaluation of the Lorentz condition. The Lorentz condition does not belong in the class of field equations, nor in the class of constitutive equations.

An admissible ad hoc method of restricting the arbitrariness in A_{κ} for waves in a medium with $\mathfrak{c}^{\nu} = 0$, is assuming $A_0 = 0$, which can always be obtained by adding a suitable multiple of the trivial solution $A_{\kappa} = k_{\kappa}$.

4. MODE-MULTIPLICITY FOR BULK PROPAGATION

It should be of some interest to investigate the different modes of propagation corresponding to solutions of the generalized wave equation (6.28). For the case of a uniform medium, with $\mathfrak{c}^{\nu} = 0$ and linear coordinates, we already found that the plane wave solution (6.33) led to the homogeneous algebraic equations of the previous section

$$\chi^{\lambda\nu\sigma\kappa}k_{\nu}k_{\sigma}\hat{A}_{\kappa} = 0\,. \tag{6.35}$$

It is known that the equation (6.35) has a nonvanishing trivial solution $A_\kappa \sim k_\kappa$, which does not correspond to a situation of physical interest. The solution $A_\kappa \sim k_\kappa$ implies that the determinant of (6.35) is identically zero for arbitrary wave vectors k_ν;

$$|\varphi^{\lambda\kappa}| = |\chi^{\lambda\nu\sigma\kappa}k_\nu k_\sigma| \equiv 0\,. \tag{6.36}$$

Any information of physical interest, therefore, may be expected to be associated with a further reduction in the rank of the determinant equation (6.36), because otherwise it would not be possible to have a solution $\hat{A}_\kappa$ not parallel to k_κ. We shall now investigate the physically nontrivial relations between frequency and wave number as well as the condition for which these lead to an unattenuated propagation in an absorption-free medium. The medium may be nonreciprocal, dispersive and anisotropic; it is assumed, however, that it is linear and uniform. In the code which was discussed in § 6.1, we propose to treat case (I, II, C, D, V, F).

The equations (6.35) and (6.36) both have the (linear) invariant form, valid for arbitrary linear coordinates. Hence there is no sacrifice of generality if we assume the frequency wave vector in the form

$$k_\nu = (k_0, k_1, 0, 0)\,, \tag{6.37}$$

which means that the coordinates have been adapted to the wave front so that $k_2 = k_3 = 0$. The equation (6.35) can then be written in the following more explicit form

$$\{\chi^{\lambda 00\kappa}k_0^2 + (\chi^{\lambda 01\kappa} + \chi^{\lambda 10\kappa})k_0 k_1 + \chi^{\lambda 11\kappa}k_1^2\}\hat{A}_\kappa = 0\,. \tag{6.38}$$

One can now introduce the ratio

$$\frac{k_0}{k_1} = u \tag{6.39}$$

as the phase velocity, because k_0 is equal to the frequency ω, and k_1, the only component of the wave vector, is equal to the wave number. The phase velocity u is the unknown of the problem.

Dividing by k_1^2, one can write (6.38) in the form

$$\{A^{\lambda\kappa}u^2 + B^{\lambda\kappa}u + C^{\lambda\kappa}\}\hat{A}_\kappa = 0\,, \tag{6.40}$$

with

$$A^{\lambda\kappa} = \chi^{\lambda 00\kappa}\,,$$

$$B^{\lambda\kappa} = \chi^{\lambda 01\kappa} + \chi^{\lambda 10\kappa}\,,$$

$$C^{\lambda\kappa} = \chi^{\lambda 11\kappa}\,.$$

A comparison with (6.21) shows that

$$A^{\lambda\kappa} = \begin{pmatrix} 0 & 0 & 0 & 0 \\ 0 & \varepsilon_{11} & \varepsilon_{12} & \varepsilon_{13} \\ 0 & \varepsilon_{12}^* & \varepsilon_{22} & \varepsilon_{23} \\ 0 & \varepsilon_{13}^* & \varepsilon_{23}^* & \varepsilon_{33} \end{pmatrix} \text{Hermitian} \tag{6.41}$$

$$B^{\lambda\kappa} = \begin{pmatrix} 0 & -\varepsilon_{11} & -\varepsilon_{12} & -\varepsilon_{13} \\ -\varepsilon_{11}^* & 0 & -\tilde{\gamma}_{13} & \tilde{\gamma}_{12} \\ -\varepsilon_{12}^* & -\tilde{\gamma}_{13}^* & -(\tilde{\gamma}_{23}+\tilde{\gamma}_{23}^*) & (\tilde{\gamma}_{22}-\tilde{\gamma}_{33}^*) \\ -\varepsilon_{13}^* & \tilde{\gamma}_{12}^* & (\tilde{\gamma}_{22}^*-\tilde{\gamma}_{33}) & (\tilde{\gamma}_{32}+\tilde{\gamma}_{32}^*) \end{pmatrix} \text{Hermitian} \tag{6.42}$$

$$C^{\lambda\kappa} = \begin{pmatrix} \varepsilon_{11} & 0 & \tilde{\gamma}_{13} & -\tilde{\gamma}_{12} \\ 0 & 0 & 0 & 0 \\ \tilde{\gamma}_{13}^* & 0 & -\chi_{33} & \chi_{23}^* \\ -\tilde{\gamma}_{12}^* & 0 & \chi_{23} & -\chi_{22} \end{pmatrix} \text{Hermitian} \tag{6.43}$$

The equation (6.40) represents an algebraic eigenvalue problem with matrix elements that have a quadratic dependence on the eigenvalue u. A contractive multiplication of (6.41) with the complex conjugate $\hat{A}^*_\lambda$ of the potential vector shows that u satisfies the following quadratic equation

$$H_1 u^2 + H_2 u + H_3 = 0 \tag{6.44}$$

with

$$H_1 = A^{\lambda\kappa}\hat{A}^*_\lambda \hat{A}_\kappa \quad \text{(positive)}$$

$$H_2 = B^{\lambda\kappa}\hat{A}^*_\lambda \hat{A}_\kappa$$

$$H_3 = C^{\lambda\kappa}\hat{A}^*_\lambda \hat{A}_\kappa \,.$$

The coefficients H_1, H_2 and H_3 of (6.44) are all real, because the matrices $A^{\lambda\kappa}$, $B^{\lambda\kappa}$ and $C^{\lambda\kappa}$ are Hermitian. Hence the eigenvalues u of (6.40) are real, if

$$H_2^2 - 4H_1H_3 \geq 0 \,. \tag{6.45}$$

The sign of the expression (6.45) does not depend on the sign of the Hermitian form H_2. It is sufficient to satisfy the condition (6.45) if H_1 and H_3 have opposite signs. An inspection of (6.41) shows that H_1 is certainly positive if the dielectric matrix is positive definite, which means that the dielectric permittivity in the directions of the principal axes is always positive. LANDAU and LIFSCHITZ [1957] give a discussion of the sign of the dielectric susceptibility in § 13 and § 14 of their book, leading to the general conclusion that the principal values of the dielectric matrix are always positive and greater than one.

A negative principal value in the present context could lead to a complex or to a purely imaginary phase velocity associated with certain directions in the medium. A direction with a purely imaginary phase velocity could formally be interpreted as a direction in which the waves are evanescent. This situation is conceivable in a three-dimensional medium with a spatial periodicity of the dielectric constant. The periodicity then leads to pass bands and stop bands for certain frequency intervals†. We may refrain, however, from discussing to what extent periodic media can be formally represented by a spatially constant but complex dielectric matrix with elements depending on frequency and wave number. It will be clear that the formal use of a negative dielectric constant does not interfere with the purely physical arguments of Landau and Lifschitz mentioned earlier in this section.

We can now continue our discussion of the condition (6.45). Having obtained the conclusion that H_1 is positive for the practical purposes in which we are interested here, we may now turn our attention to the sign of H_3. An inspection of the corresponding matrix (6.43) shows that the arguments previously used for (6.41) do not apply to (6.43), because permeability and permittivity, and also cross terms $\tilde{\gamma}_{lk}$, occur simultaneously in H_3. It is possible, however, to reduce the discussion of the sign H_3 to a discussion of the permeability part, i.e.,

$$\begin{pmatrix} -\chi_{33} & \overset{*}{\chi}_{23} \\ \chi_{23} & -\chi_{22} \end{pmatrix}, \tag{6.46}$$

if we use the gauge freedom in the A_κ so that $A_0 = 0$††. We may now assume on similar grounds as for the dielectric matrix that the permeability matrix has positive principal values only. The same is then true for the inverse of the

† Compare L. BRILLOUIN [1947] § 35.

†† Compare the concluding remarks of § 6.3.

permeability matrix i.e. χ_{lk} and for our present purpose it is true for the submatrix

$$\begin{pmatrix} \chi_{33} & \chi_{23}^* \\ \chi_{23} & \chi_{22} \end{pmatrix}. \tag{6.47}$$

If (6.47) has positive principal values, then (6.46) has negative principal values, because the products of the roots of their characteristic equations are the same for (6.46) and (6.47), but the sums of the roots have the opposite sign. Hence H_3 is negative, consequently the condition (6.45) for the reality of the roots of (6.44) is satisfied under the circumstances stipulated above. It will be clear that the previous arguments apply to the nontrivial roots, because the gauge assumption $A_0 = 0$ has already eliminated the occurrence of a trivial solution. It should be noted that this same gauge condition causes H_2 to vanish for a medium at rest, because only the imaginary anti-symmetric terms (optical activity) then appear in (6.42). Hence for an absorption-free medium at rest we may conclude from (6.44) that the phase velocity is either real or imaginary depending on the sign of the principal values of the permittivity and permeability matrices.

The fact that H_2 may vanish, notwithstanding a $\tilde{\gamma}_{lk} \neq 0$, does not imply that the phase velocities do not depend on the elements of $\tilde{\gamma}_{lk}$. The equation (6.44) of course is only a means to investigate the nature of the roots u and does not provide a basis for their computation, because the coefficients H_1, H_2 and H_3 are still functions of the eigenvector solutions $\hat{A}_\kappa$.

A computation of the eigenvalues u starts from the degenerate determinant equation (6.36). This determinant, for the case under discussion as given by equation (6.40), takes the form

$$|A^{\lambda\kappa}u^2 + B^{\lambda\kappa}u + C^{\lambda\kappa}| \equiv 0. \tag{6.48}$$

The fact that it vanishes identically can be easily checked by writing out the determinant using the explicit forms given by (6.41), (6.42) and (6.43). One finds that the first two rows and columns of (6.48) have a proportionality factor u. The degeneracy of (6.48) ties in with the fact that the gauge assumption $A_0 = 0$ reduces the number of variables to three. We are thus left with three equations for the three spatial components of the potential vector. The compatibility of these equations requires that the corresponding subdeterminant of (6.48) should vanish, i.e.

$$\begin{vmatrix} \varepsilon_{11} & \{-\varepsilon_{12}u+\tilde{\gamma}_{13}\} & \{-\varepsilon_{13}u-\tilde{\gamma}_{12}\} \\ \{-\varepsilon_{12}^{*}u+\tilde{\gamma}_{13}^{*}\} & \{\varepsilon_{22}u^2-(\tilde{\gamma}_{23}+\tilde{\gamma}_{23}^{*})u-\chi_{33}\} & \{\varepsilon_{23}u^2+(\tilde{\gamma}_{22}-\tilde{\gamma}_{33}^{*})u+\chi_{23}^{*}\} \\ \{-\varepsilon_{13}^{*}u-\tilde{\gamma}_{12}^{*}\} & \{\varepsilon_{23}^{*}u^2+(\tilde{\gamma}_{22}^{*}-\tilde{\gamma}_{33})u+\chi_{23}\} & \{\varepsilon_{33}u^2+(\tilde{\gamma}_{32}+\tilde{\gamma}_{32}^{*})u-\chi_{22}\} \end{vmatrix} = 0. \tag{6.49}$$

Note that (6.49) is a sufficient condition to reduce the rank of (6.48), because all the other three-by-three subdeterminants are already zero. Hence (6.49) characterizes an invariant feature of (6.40). An inspection shows that (6.49) is of degree 4 in the phase velocity u, which means that a given orientation of the wave front has at most four distinct propagation velocities associated with it. One might have sensed this as a general result by realizing that a moving birefringent medium indeed produces four different phase velocities for a given orientation of the wavefront, if one applies the Fresnel-Fizeau drag formula to the well-known results for the stationary birefringent medium. It is easy to check that a root symmetry for the opposite sense of direction in the medium at rest occurs only if the Faraday effect is absent. Special cases are discussed in § 8.2 and § 8.3.

CHAPTER 7

NONUNIFORMITY, ANISOTROPY AND GENERAL COORDINATES IN THREE-SPACE

1. THE COVARIANT D'ALEMBERTIANS IN THREE-SPACE

In the previous chapter we derived a generalized wave equation (6.28) for the electromagnetic potentials. It was stated that this wave equation was found invariant for general space-time transformation and that it could be applied to media with nonuniform, nonreciprocal and anisotropic properties. In the present chapter we want to examine in particular the covariance with respect to the subset of general space transformations. We shall derive the three dimensionally covariant wave equations for the electric and magnetic field vectors. The equations thus obtained should give us an opportunity to investigate how these generalized forms tie in with the more familiar and conventional presentations of electromagnetic theory. We want to show in particular the advantage of the natural invariant forms for writing down explicit equations on specific curvilinear coordinates. As regards the physical features of nonuniformity and anisotropy, only the fundamental and basic equations will be discussed. We shall not attempt here to establish the general physical aspects of wave propagation in nonuniform and anisotropic bodies. It will be seen that the formulation of the fundamental equations is interesting enough in itself to justify a separate discussion.

Possibly one of the most striking features of the natural invariant representation with respect to customary presentations is that the equations can be used in the same form in any arbitrary system of curvilinear coordinates. It is quite amazing that one has refrained so long from using these relatively well established mathematical techniques in electromagnetics, because it requires only minor modifications in the definitions of the components of the physical fields with respect to curvilinear coordinates. In point of fact, these very same modifications have been well accepted for more than a century in the formalism of analytical dynamics. In analytical dynamics it is common procedure to associate an angular momentum with an angular coordinate, so that each of them has a physical dimension different from the linear coordinate and the linear momentum. The dimension of the product of these two variables, however, is not affected. The rules given in chapter 2 account for these differences in physical dimensions between the components of a field and lead quite naturally to the fact that the momentum vector should be regarded as a covariant vector.

The customary treatment of curvilinear coordinates in electromagnetics and continuum mechanics, however, is entirely different. The standard procedure is there, to introduce a local system of Cartesian frames tangent to the curvilinear frame. This has the advantage that the conventional rules of Cartesian vectors and tensor algebra remain valid in these local frames. Moreover the components of the field vectors all maintain the same physical dimension in these local frames of reference. However, difficulties with the usual Cartesian expressions occur as soon as one considers differential operations in these local frames, because the orientations of the local frames are functions of position. Differential operations then invoke a set of complex correction terms which are absent in the simpler form invariant representations. It was mentioned already in § 3.2 that the naturally invariant divergence and curl operators are not valid in anholonomic frames of reference. The reader who is familiar with this somewhat more specialized aspect of the absolute differential calculus may recognize the procedure with local Cartesian frames as a case in point where use is being made of an anholonomic system of reference in field theory†.

The question now arises: Why use an anholonomic system of reference if a holonomic frame is readily available? In mechanics there are situations (billiard ball on a rough table) with nonintegrable conditions of constraint, the anholonomity is then an essential feature of the problem. This is not the case in the present field theoretical problem, provided one knows how to extend the vector concept from Cartesian to curvilinear systems. Einstein may have been the first to stress the use of the generalized vector concept in curvilinear systems for physics. The 19th century originator of our customary curl and divergence expressions for curvilinear systems, however, may have had reasons not to commit himself with the transition of the vector concept from Cartesian to non-Cartesian systems. He found an ingenious way of evading the issue by introducing his local Cartesian frames, which we now recognize as having anholonomic properties.

Rather than reiterate this typical example of classical analysis from here to eternity, we may compare instead its virtues with two other methods of approach for obtaining vector differential expressions on curvilinear systems. The three major methods of approach which we like to consider are:

(1) local Cartesian frames	anholonomic frames Cartesian vector concept
(2) covariant derivatives (3) natural derivatives	holonomic frames, but require the vector concept for curvilinear systems

† For the reader who likes to check this, please compare the curl and divergence expressions for anholonomic coordinates given by SCHOUTEN [1951] p. 103 formulae 7.4a and 7.4b. An application of these expressions, which requires a computation of the object of anholonomity, leads to the conventional curl and divergence expressions for curvilinear coordinates listed in any text book of physics.

The vector concept for curvilinear systems leads of course to nothing else than the distinction between covariant vectors, contravariant vectors and vector densities, as elaborately discussed in the previous chapters. The presence of the metric tensor, however, enables one to make any desired transition between any of these different vector representations. A transition to the appropriate vector representation will (in electromagnetism) reduce method (2) to method (3), because the terms with the Christoffel symbols then cancel for the divergence and curl operators. We verified in chapter 3 that this situation arises for the fundamental equations of electromagnetics if one adheres to what might be called the "intrinsic" representation of the fields discussed in chapter 2.

Summarizing our conclusions we find that:

(1) requires a calculation of the object of anholonomity

(2) requires a calculation of the Christoffel symbols

(3) can be applied directly, without the use of any non-invariant correction terms.

A transition from (1) to (3) or from (1) to (2) requires a modification of one or more vector components to account for the transition from local Cartesian to curvilinear components. The reader may compare the equations (7.47) and (7.48) in this chapter; they denote the difference between the angular components for cylindrical coordinates of the electric field vector in curvilinear and in local Cartesian coordinates.

It will be clear that one can make almost any number of analytical exercises exploring the relationship between the methods (1), (2) and (3). In the literature one finds almost exclusively a shorthand version of method (1). Method (2) is frequently discussed, but seldomly applied (except for the Laplacian of scalars) because like method (3) it requires a generalization of the Cartesian vector concept. Method (3), being the simplest, will now be treated in this chapter for the three-space dimensions. We shall verify the correspondence with method (1) after simplifying the results for uniform and isotropic media.

In § 3.4 we obtained a natural invariant representation of the Maxwell equations for three dimensions *valid in any spatial curvilinear coordinate system*:

$$2\partial_{[\lambda}E_{\nu]} = -\dot{B}_{\lambda\nu}\,; \tag{7.1}$$

$$\partial_{[\kappa}B_{\lambda\nu]} = 0\,; \tag{7.2}$$

$$\partial_{\nu}\mathfrak{H}^{\lambda\nu} = \dot{\mathfrak{D}}^{\lambda} + \mathfrak{c}^{\lambda}\,; \tag{7.3}$$

$$\partial_{\lambda}\mathfrak{D}^{\lambda} = \rho\,, \tag{7.4}$$

with

$$B_{\lambda\nu} = -B_{\nu\lambda} \tag{7.5}$$

and

$$\mathfrak{H}^{\lambda\nu} = -\mathfrak{H}^{\nu\lambda}, \quad \lambda, \nu, \ldots = 1, 2, 3\,. \tag{7.6}$$

The invariance holds if the following identification is adopted for a Cartesian frame

$$\begin{aligned} &E_1 = E_x\,; \quad B_{23} = B_x\,; \quad \mathfrak{D}^1 = D_x\,; \quad \mathfrak{H}^{23} = H_x\,; \quad \mathfrak{c}^1 = s_x\,; \\ &E_2 = E_y\,; \quad B_{31} = B_y\,; \quad \mathfrak{D}^2 = D_y\,; \quad \mathfrak{H}^{31} = H_y\,; \quad \mathfrak{c}^2 = s_y\,; \\ &E_3 = E_z\,; \quad B_{12} = B_z\,; \quad \mathfrak{D}^3 = D_z\,; \quad \mathfrak{H}^{12} = H_z\,; \quad \mathfrak{c}^3 = s_z\,; \end{aligned} \tag{7.7}$$

$$\rho = \rho\,,$$

and provided the field quantities obey the following transformation rules for a change in coordinates:

$$E_{\lambda'} = A^{\lambda}_{\lambda'}E_{\lambda}\,; \tag{7.8}$$

$$\mathfrak{D}^{\lambda'} = |\Delta|^{-1}A^{\lambda'}_{\lambda}\mathfrak{D}^{\lambda}\,; \tag{7.9}$$

$$\mathfrak{c}^{\lambda'} = |\varDelta|^{-1}A_{\lambda}^{\lambda'}\mathfrak{c}^{\lambda}; \tag{7.10}$$

$$B_{\lambda'\nu'} = A_{\lambda'}^{\lambda}A_{\nu'}^{\nu}B_{\lambda\nu}; \tag{7.11}$$

$$\mathfrak{H}^{\lambda'\nu'} = |\varDelta|^{-1}A_{\lambda}^{\lambda'}A_{\nu}^{\nu'}\mathfrak{H}^{\lambda\nu}; \tag{7.12}$$

$$\rho(\kappa') = |\varDelta|^{-1}\rho(\kappa), \tag{7.13}$$

with

$$A_{\lambda'}^{\lambda} = \frac{\partial x^{\lambda}}{\partial x^{\lambda'}}; \quad A_{\lambda}^{\lambda'} = \frac{\partial x^{\lambda'}}{\partial x^{\lambda}}; \quad \varDelta = |A_{\lambda}^{\lambda'}| \quad \text{(Jacobian determinant)}. \tag{7.14}$$

The dielectric behavior of a linear, anisotropic, nondispersive medium is given by

$$\mathfrak{D}^{\lambda} = \varepsilon^{\lambda\nu}E_{\nu}. \tag{7.15}$$

The transformation of $\mathfrak{D}^{\lambda}$ and E_{λ} is fixed by (7.8) and (7.9), so the transformation of $\varepsilon^{\lambda\nu}$ must be the transformation of a contravariant, symmetric tensor density of valence two and of weight $+1$,

$$\varepsilon^{\lambda'\nu'} = |\varDelta|^{-1}A_{\lambda}^{\lambda'}A_{\nu}^{\nu'}\varepsilon^{\lambda\nu}. \tag{7.16}$$

The corresponding relation for the magnetic behavior of a linear nondispersive medium is

$$B_{\lambda\nu} = \tfrac{1}{2}\mu_{\lambda\nu\sigma\kappa}\mathfrak{H}^{\sigma\kappa}. \tag{7.17}$$

The medium is nonuniform if $\varepsilon^{\lambda\nu}$ and $\mu_{\lambda\nu\sigma\kappa}$ are still functions of the coordinates on a linear (uniform) frame of reference. The transformation formulae (7.11) and (7.12) determine the transformation properties of $\mu_{\lambda\nu\sigma\kappa}$ as that of a tensor density of weight -1

$$\mu_{\lambda'\nu'\sigma'\kappa'} = |\varDelta|A_{\lambda'}^{\lambda}A_{\nu'}^{\nu}A_{\sigma'}^{\sigma}A_{\kappa'}^{\kappa}\mu_{\lambda\nu\sigma\kappa}, \tag{7.18}$$

with the symmetries

$$\mu_{\lambda\nu\sigma\kappa} = \mu_{\sigma\kappa\lambda\nu}, \tag{7.19}$$

$$\mu_{\lambda\nu\sigma\kappa} = -\mu_{\nu\lambda\sigma\kappa}. \tag{7.20}$$

For reasons that will become obvious it turns out to be convenient to use the inverse of (7.17)

$$\mathfrak{H}^{\lambda\nu} = \tfrac{1}{2}\chi^{\lambda\nu\sigma\kappa}B_{\sigma\kappa}. \tag{7.21}$$

The $\chi^{\lambda\nu\sigma\kappa}$ now is a density of weight $+1$ with transformation

$$\chi^{\lambda'\nu'\sigma'\kappa'} = |\Delta|^{-1}A_{\lambda}^{\lambda'}A_{\nu}^{\nu'}A_{\sigma}^{\sigma'}A_{\kappa}^{\kappa'}\chi^{\lambda\nu\sigma\kappa}, \tag{7.22}$$

and it has the same symmetries as $\mu_{\lambda\nu\sigma\kappa}$

$$\chi^{\lambda\nu\sigma\kappa} = \chi^{\sigma\kappa\lambda\nu} = -\chi^{\nu\lambda\sigma\kappa}. \tag{7.23}$$

The wave equations for the components of the electrical field are derived by a substitution of (7.21) into (7.3)

$$\tfrac{1}{2}\partial_{\nu}\chi^{\lambda\nu\sigma\kappa}B_{\sigma\kappa} = \dot{\mathfrak{D}}^{\lambda} + \mathfrak{c}^{\lambda}. \tag{7.24}$$

Differentiation with respect to the time yields, if one assumes that $\mathfrak{c}^{\lambda} = 0$ and that $\chi^{\lambda\nu\sigma\kappa}$ is independent of time,

$$\tfrac{1}{2}\partial_{\nu}\chi^{\lambda\nu\sigma\kappa}\dot{B}_{\sigma\kappa} = \ddot{\mathfrak{D}}^{\nu}. \tag{7.25}$$

Now one uses (7.1) and (7.15) and one also assumes that $\varepsilon^{\lambda\nu}$ is independent of time, then

$$\boxed{\partial_{\nu}\chi^{\lambda\nu\sigma\kappa}\partial_{\sigma}E_{\kappa} = -\varepsilon^{\lambda\mu}\ddot{E}_{\mu}} \tag{7.26}$$

These are a set of simultaneous wave equations with an

accessory condition (7.4) with $\rho = 0$. Considering (7.15) one obtains

$$\boxed{\partial_\lambda \varepsilon^{\lambda\nu} E_\nu = 0} \tag{7.27}$$

The three equations (7.26) and the equation (7.27) are valid in any curvilinear coordinate system. The medium does not have to be uniform; the only restrictions are linearity and local interaction as expressed by (7.15) and (7.21) and independence of $\varepsilon^{\lambda\nu}$ and $\chi^{\lambda\nu\sigma\kappa}$ of the time.

It should be noted that (7.26) and (7.27) are dependent. The divergence of (7.26) gives a zero left-hand member, hence $\partial_\lambda \varepsilon^{\lambda\mu} \ddot{E}_\mu = 0$.

It is useful to check whether (7.26) and (7.27) reduce to the usual d'Alembertian on a Cartesian frame for conditions of isotropy and uniformity of the medium†.

For isotropy the permittivity and permeability matrices reduce to

$\varepsilon^{\lambda\nu}$	1	2	3
1	ε	0	0
2	0	ε	0
3	0	0	ε

$\chi^{\lambda\nu\sigma\kappa}$	23	31	12
23	$1/\mu$	0	0
31	0	$1/\mu$	0
12	0	0	$1/\mu$

The expansion of (7.26) yields for $\lambda = 1$ as the only nonvanishing terms,

$$\chi^{1212}\partial_2\partial_1 E_2 + \chi^{1221}\partial_2\partial_2 E_1 + \chi^{1313}\partial_3\partial_1 E_3 + \chi^{1331}\partial_3\partial_3 E_1 = -\varepsilon^{11}\ddot{E}_1\,,$$

† A uniform medium is defined as a medium having a constant $\varepsilon^{\lambda\nu}$ and $\chi^{\lambda\nu\sigma\kappa}$ if referred to a Cartesian frame.

and (7.27) is

$$\varepsilon^{11}\partial_1 E_1 + \varepsilon^{22}\partial_2 E_2 + \varepsilon^{33}\partial_3 E_3 = 0\,,$$

because the χ and ε are now constants in view of the condition of uniformity. Using the tables for ε and χ one obtains

$$-\,\partial_1(\partial_2 E_2 + \partial_3 E_3) + \partial_2\partial_2 E_1 + \partial_3\partial_3 E_1 = \varepsilon\mu\ddot{E}_1\,,$$

and

$$\partial_1 E_1 + \partial_2 E_2 + \partial_3 E_3 = 0\,.$$

The first term of the first equation can be converted into $\partial_1\partial_1 E_1$ by means of the second equation. Thus one obtains the conventional d'Alembertian

$$(\partial_1\partial_1 + \partial_2\partial_2 + \partial_3\partial_3)E_1 = \varepsilon\mu\ddot{E}_1\,,$$

or

$$\nabla^2 E_x = \varepsilon\mu\ddot{E}_x\,.$$

The equations for the other components are obtained by choosing $\lambda = 2$ and $\lambda = 3$, and as may be expected they lead to the same d'Alembertian.

For the components of the other field quantities a similar elimination procedure can be followed as just demonstrated for the E field leading to (7.26) and (7.27). Similarly, all these systems analogous to (7.26) and (7.27) reduce to the same d'Alembertian in a Cartesian frame if the medium is uniform and isotropic. The complete set is given below, where $\eta_{\lambda\nu}$ denotes the reciprocal of $\varepsilon^{\lambda\nu}$:

E	$\partial_\nu\chi^{\lambda\nu\sigma\kappa}\partial_\sigma E_\kappa + \varepsilon^{\lambda\mu}\ddot{E}_\mu = 0$	$\partial_\lambda\varepsilon^{\lambda\nu}E_\nu = 0$
B	$\partial_{[\kappa}\eta_{\sigma]\lambda}\partial_\nu\chi^{\lambda\nu\mu\rho}B_{\mu\rho} + \ddot{B}_{\kappa\sigma} = 0$	$\partial_{[\kappa}B_{\lambda\nu]} = 0$
D	$\partial_\nu\chi^{\lambda\nu\sigma\kappa}\partial_\sigma\eta_{\kappa\mu}\mathfrak{D}^\mu + \ddot{\mathfrak{D}}^\lambda = 0$	$\partial_\lambda\mathfrak{D}^\lambda = 0$
H	$\partial_{[\kappa}\eta_{\sigma]\lambda}\partial_\nu\mathfrak{H}^{\lambda\nu} + \frac{1}{4}\mu_{\kappa\sigma\mu\rho}\ddot{\mathfrak{H}}^{\mu\rho} = 0$	$\partial_{[\kappa}\mu_{\lambda\nu]\sigma\tau}\mathfrak{H}^{\sigma\tau} = 0$

These equations are valid in any coordinate system if the E_λ and $B_{\lambda\nu}$ transform as ordinary (nondensity) quantities, the $\mathfrak{H}^{\lambda\nu}$, $\mathfrak{D}^\nu$, $\varepsilon^{\lambda\nu}$ and $\chi^{\lambda\nu\sigma\kappa}$ transform as densities of weight $+1$ and the $\mu_{\lambda\nu\sigma\kappa}$ and $\eta_{\lambda\nu}$ transform as densities of weight -1. Provided we abide by the transformation rules, these equations can be applied to any linear, anisotropic, nondispersive, and nonuniform medium and they can be specified for any coordinate system that happens to be of interest. In the next section the case of a cylindrically symmetric medium in cylindrical coordinates will be considered as an example.

2. THE WAVE EQUATIONS FOR A MEDIUM WITH CYLINDRICAL SYMMETRY IN CYLINDRICAL COORDINATES

The transformation equations for a transition from orthogonal Cartesian coordinates to cylindrical coordinates are known to be

$$\begin{aligned} x &= r\cos\varphi\,; \\ y &= r\sin\varphi\,; \\ z &= z\,. \end{aligned} \tag{7.28}$$

The x, y and z are labeled by 1, 2, 3 and r, φ, z by $1'$, $2'$, $3'$. The Jacobian transformation matrices then become

$$\begin{pmatrix} \dfrac{\partial r}{\partial x} & \dfrac{\partial r}{\partial y} & \dfrac{\partial r}{\partial z} \\ \dfrac{\partial \varphi}{\partial x} & \dfrac{\partial \varphi}{\partial y} & \dfrac{\partial \varphi}{\partial z} \\ \dfrac{\partial z}{\partial x} & \dfrac{\partial z}{\partial y} & \dfrac{\partial z}{\partial z} \end{pmatrix} = \begin{pmatrix} A_1^{1'} & A_2^{1'} & A_3^{1'} \\ A_1^{2'} & A_2^{2'} & A_3^{2'} \\ A_1^{3'} & A_2^{3'} & A_3^{3'} \end{pmatrix} = \begin{pmatrix} \cos\varphi & \sin\varphi & 0 \\ -\dfrac{\sin\varphi}{r} & \dfrac{\cos\varphi}{r} & 0 \\ 0 & 0 & 1 \end{pmatrix}, \tag{7.29}$$

and the inverse

$$\begin{pmatrix} \frac{\partial x}{\partial r} & \frac{\partial x}{\partial \varphi} & \frac{\partial x}{\partial z} \\ \frac{\partial y}{\partial r} & \frac{\partial y}{\partial \varphi} & \frac{\partial y}{\partial z} \\ \frac{\partial z}{\partial r} & \frac{\partial z}{\partial \varphi} & \frac{\partial z}{\partial z} \end{pmatrix} = \begin{pmatrix} A^1_{1'} & A^1_{2'} & A^1_{3'} \\ A^2_{1'} & A^2_{2'} & A^2_{3'} \\ A^3_{1'} & A^3_{2'} & A^3_{3'} \end{pmatrix} = \begin{pmatrix} \cos\varphi & -r\sin\varphi & 0 \\ \sin\varphi & r\cos\varphi & 0 \\ 0 & 0 & 1 \end{pmatrix}. \quad (7.30)$$

The Jacobian determinants are

$$\Delta = |A^{\lambda'}_{\lambda}| = r^{-1} \quad \text{and} \quad |A^{\lambda}_{\lambda'}| = \Delta^{-1} = r. \quad (7.31)$$

We will assume the medium to have rotational symmetry around the z-axis. In a Cartesian system the tables for $\chi^{\lambda\nu\sigma\kappa}$ and $\varepsilon^{\lambda\nu}$ then are

$\varepsilon^{\lambda\nu}$	1	2	3
1	ε_t	0	0
2	0	ε_t	0
3	0	0	ε_a

, (7.32)

$\chi^{\lambda\nu\sigma\kappa}$	23	31	12
23	χ_t	0	0
31	0	χ_t	0
12	0	0	χ_a

. (7.33)

The ε_t is the transverse dielectric permittivity and χ_t is the reciprocal of the transverse magnetic permeability: both of them may be assumed to be functions of r, i.e., the nonuniformity is given as a tapering of dielectric and magnetic density in the radial direction. The ε_a and χ_a are the corresponding quantities in the (axial) z-direction: they too may be functions of r only. The ratios $\varepsilon_a/\varepsilon_t$ and χ_a/χ_t determine the electric and magnetic anisotropy.

To obtain the analogues of tables (7.32) and (7.33) for a cylindrical frame of reference, the transformation formulae (7.16) and (7.22) have to be used. For $\varepsilon^{1'1'}$, for instance, we get

$$\varepsilon^{1'1'} = \Delta^{-1}(A_1^{1'}A_1^{1'}\varepsilon^{11} + A_2^{1'}A_2^{1'}\varepsilon^{22} + A_3^{1'}A_3^{1'}\varepsilon^{33}) .$$

Using (7.29), (7.30), (7.31) and (7.32) this becomes

$$\varepsilon^{1'1'} = r(\varepsilon_t \cos^2\varphi + \varepsilon_t \sin^2\varphi) = r\varepsilon_t .$$

The complete result is

$\varepsilon^{\lambda'\nu'}$	$1'$	$2'$	$3'$
$1'$	$r\varepsilon_t$	0	0
$2'$	0	$r^{-1}\varepsilon_t$	0
$3'$	0	0	$r\varepsilon_a$

(7.34)

$\chi^{\lambda'\nu'\sigma'\kappa'}$	$2'3'$	$3'1'$	$1'2'$
$2'3'$	$r^{-1}\chi_t$	0	0
$3'1'$	0	$r\chi_t$	0
$1'2'$	0	0	$r^{-1}\chi_a$

(7.35)

It is not necessary to work out the complex transformation (7.22) for $\chi^{\lambda\nu\sigma\kappa}$. The antisymmetry in $\lambda\nu$ and $\sigma\kappa$ enables us to use the duality transcription $12 \to 3$ etc. The resulting tensor of valence 2 then should be regarded as a double covariant tensor density of weight -1. It may be verified that the result of this transformation is identical with (7.22).

The tables (7.34) and (7.35) provide all the necessary material to expand the covariant wave equations to write them in explicit form for cylindrical coordinates. We may take the system for the electrical vector and proceed along the same lines as at the end of the previous section. The equations on cylindrical coordinates are (7.26) and (7.27) with $\varepsilon^{\lambda\nu}$ and $\chi^{\lambda\nu\sigma\kappa}$ substituted according to tables (7.34) and (7.35). The general procedure requires the expansion of

$$\partial_{\nu'}\chi^{\lambda'\nu'\sigma'\kappa'}\partial_{\sigma'}E_{\kappa'} + \varepsilon^{\lambda'\mu'}\ddot{E}_{\kappa'} = 0\,; \tag{7.26}$$

$$\partial_{\lambda'}\varepsilon^{\lambda'\nu'}E_{\nu'} = 0\,. \tag{7.27}$$

The primes are meant to indicate that the coordinate system is the r, φ, z system. Expanding (keeping in mind the diagonal properties of (7.34) and (7.35)), one obtains for $\lambda = 1, 2, 3$

$$\begin{aligned}
&\lambda = 1 \begin{cases} \partial_2\chi^{1212}\partial_1E_2 + \partial_2\chi^{1221}\partial_2E_1 + \\ \partial_3\chi^{1313}\partial_1E_3 + \partial_3\chi^{1331}\partial_3E_1 + \varepsilon^{11}\ddot{E}_1 = 0 \end{cases} \\
&\lambda = 2 \begin{cases} \partial_1\chi^{2121}\partial_2E_1 + \partial_1\chi^{2112}\partial_1E_2 + \\ \partial_3\chi^{2323}\partial_2E_3 + \partial_3\chi^{2332}\partial_3E_2 + \varepsilon^{22}\ddot{E}_2 = 0 \end{cases} \\
&\lambda = 3 \begin{cases} \partial_1\chi^{3131}\partial_3E_1 + \partial_1\chi^{3113}\partial_1E_3 + \\ \partial_2\chi^{3232}\partial_3E_2 + \partial_2\chi^{3223}\partial_2E_3 + \varepsilon^{33}\ddot{E}_3 = 0 \end{cases}
\end{aligned} \tag{7.36}$$

$$\partial_1\varepsilon^{11}E_1 + \partial_2\varepsilon^{22}E_2 + \partial_3\varepsilon^{33}E_3 = 0\,. \tag{7.37}$$

The primes have now been omitted for the sake of simplicity, the transcription $1 \to r$, $2 \to \varphi$, $3 \to z$ and the tables (7.34) and (7.35) lead to the following forms:

$$\left\{\begin{aligned}&\frac{\partial}{\partial\varphi} r^{-1}\chi_{\mathrm{a}}\frac{\partial}{\partial r}E_{\varphi} - \frac{\partial}{\partial\varphi} r^{-1}\chi_{\mathrm{a}}\frac{\partial}{\partial\varphi}E_r + \\ &\frac{\partial}{\partial z} r\chi_{\mathrm{t}}\frac{\partial}{\partial r}E_z - \frac{\partial}{\partial z} r\chi_{\mathrm{t}}\frac{\partial}{\partial z}E_r + r\varepsilon_{\mathrm{t}}\ddot{E}_r = 0\end{aligned}\right. \tag{7.38}$$

$$\left\{\begin{aligned}&\frac{\partial}{\partial r} r^{-1}\chi_{\mathrm{a}}\frac{\partial}{\partial\varphi}E_r - \frac{\partial}{\partial r} r^{-1}\chi_{\mathrm{a}}\frac{\partial}{\partial r}E_{\varphi} + \\ &\frac{\partial}{\partial z} r^{-1}\chi_{\mathrm{t}}\frac{\partial}{\partial\varphi}E_z - \frac{\partial}{\partial z} r^{-1}\chi_{\mathrm{t}}\frac{\partial}{\partial z}E_{\varphi} + r^{-1}\varepsilon_{\mathrm{t}}\ddot{E}_{\varphi} = 0\end{aligned}\right. \tag{7.39}$$

$$\left\{\begin{aligned}&\frac{\partial}{\partial r} r\chi_{\mathrm{t}}\frac{\partial}{\partial z}E_r - \frac{\partial}{\partial r} r\chi_{\mathrm{t}}\frac{\partial}{\partial r}E_z + \\ &\frac{\partial}{\partial\varphi} r^{-1}\chi_{\mathrm{t}}\frac{\partial}{\partial z}E_{\varphi} - \frac{\partial}{\partial\varphi} r^{-1}\chi_{\mathrm{t}}\frac{\partial}{\partial\varphi}E_z + r\varepsilon_{\mathrm{a}}\ddot{E}_z = 0\end{aligned}\right. \tag{7.40}$$

$$\frac{\partial}{\partial r} r\varepsilon_{\mathrm{t}}E_r + \frac{\partial}{\partial\varphi} r^{-1}\varepsilon_{\mathrm{t}}E_{\varphi} + \frac{\partial}{\partial z} r\varepsilon_{\mathrm{a}}E_z = 0\,. \tag{7.41}$$

The equations (7.38) . . . (7.41) are the explicit forms of the wave equations for a cylindrically nonuniform, anisotropic body on cylindrical coordinates.

For isotropy $\varepsilon_{\mathrm{t}} = \varepsilon_{\mathrm{a}} = \varepsilon$, and $\chi_{\mathrm{t}} = \chi_{\mathrm{a}} = 1/\mu$. Uniformity makes ε and μ constants. One can then verify that (7.38) . . . (7.41) reduce to

$$r^{-1}\frac{\partial}{\partial r} r\frac{\partial}{\partial r}E_r + r^{-2}\frac{\partial^2}{\partial\varphi^2}E_r + \frac{\partial^2}{\partial z^2}E_r - \varepsilon\mu\ddot{E}_r = \frac{E_r}{r^2} + \frac{2}{r^3}\frac{\partial}{\partial\varphi}E_{\varphi}\,, \tag{7.42}$$

$$r^{-1}\frac{\partial}{\partial r}r\frac{\partial}{\partial r}E_\varphi + r^{-2}\frac{\partial^2}{\partial \varphi^2}E_\varphi + \frac{\partial^2}{\partial z^2}E_\varphi - \varepsilon\mu\ddot{E}_\varphi$$

$$= \frac{2}{r}\left(\frac{\partial E_\varphi}{\partial r} - \frac{\partial E_r}{\partial \varphi}\right), \quad (7.43)$$

$$r^{-1}\frac{\partial}{\partial r}r\frac{\partial}{\partial r}E_z + r^{-2}\frac{\partial^2}{\partial \varphi^2}E_z + \frac{\partial^2}{\partial z^2}E_z - \varepsilon\mu\ddot{E}_z = 0\,. \quad (7.44)$$

The component E_z has been eliminated from (7.38) and (7.39) by means of (7.41). In equation (7.40) E_r and E_φ both can be eliminated with the help of (7.41).

The left-hand members of (7.42), (7.43) and (7.44) are the conventional d'Alembertian on cylindrical coordinates as if it were operating on a scalar function. The equations (7.42) and (7.43) have a right-hand member which might be said to be due to the "geometric anisotropy" of the cylindrical coordinate system.

It should be noted that the right-hand member of (7.43) is related to the magnetic induction in the z-direction, because

$$\frac{\partial E_\varphi}{\partial r} - \frac{\partial E_r}{\partial \varphi} = 2\partial_{[1'}E_{2']} \quad (7.45)$$

is the covariant expression for the curl of the electric field in the 3′ direction.

The right-hand members of (7.42) and (7.43) are slightly different from those quoted in the literature. This is due to the difference in definition of E_φ. The E_λ, interpreted as a covariant vector, transforms according to the formula

$$E_{\lambda'} = A^{\lambda}_{\lambda'}E_\lambda\,. \quad (7.46)$$

From (7.30) we obtain

$$E_\varphi = -\,rE_x \sin\varphi + rE_y \cos\varphi\,. \quad (7.47)$$

The customary local Cartesian definition of E_φ, however, is

$$\overset{c}{E}_\varphi = - E_x \sin\varphi + E_y \cos\varphi \,. \tag{7.48}$$

The substitution of $\overset{c}{E}_\varphi = r^{-1} E_\varphi$ in (7.42) and (7.43) yields the equations in their conventional form

$$\frac{E_r}{r^2} + \frac{2}{r^2}\frac{\partial}{\partial\varphi}\overset{c}{E}_\varphi = \text{"cylindrical" d'Alembertian operating on } E_r\,, \tag{7.42a}$$

$$\frac{\overset{c}{E}_\varphi}{r^2} - \frac{2}{r^2}\frac{\partial}{\partial\varphi}E_r = \text{"cylindrical" d'Alembertian operating on } \overset{c}{E}_\varphi\,. \tag{7.43a}$$

The equations (7.42a) and (7.43a) are identical to those quoted in the textbooks (MORSE and FESHBACH [1953]).

CHAPTER 8

RECIPROCITY AND NONRECIPROCITY

1. GENERAL

The difference between reciprocity and nonreciprocity is commonly demonstrated and defined in systems which have propagation asymmetry if one interchanges signal source and signal detector. Instead of following this procedure, most common in network theory, we may consider here the possibility of defining reciprocity as a local property of the medium. For a linear medium one should be able to pin down reciprocity in the behavior of the coefficients ε_{lk}, χ_{lk} and $\tilde{\gamma}_{lk}$, the elements of the constitutive tensor (6.21). It is a common procedure to specify the spatial properties of these elements, in the sense of the Neumann principle, for the symmetry groups of crystals. ZOCHER and TÖRÖK [1955] have suggested the use of time transformations to characterize in further detail the elements of the constitutive tensor. The time transformation which one expects to play a role in the criteria for reciprocity is the simple time reversal

$$t \rightarrow -t\,. \tag{8.1}$$

In a linear Lorentz frame, time and space coordinates may be considered as the components of a contravariant vector, while frequency and the three-dimensional wave vector form a covariant four vector. Their scalar product

$$S = \omega t + k_\nu x^\nu\,, \quad \nu = 1, 2, 3\,, \tag{8.2}$$

is known as the phase. If one wants the phase to be a true scalar invariant then a time reversal (8.1) should be accompanied by a frequency reversal

$$\omega \to -\omega . \tag{8.3}$$

An application of the transformations (8.1) and (8.3) to the elements of the constitutive tensor will give us an opportunity to separate the reciprocal and nonreciprocal effects if we accept the following definition to separate reciprocity and nonreciprocity:

The elements of the constitutive tensor of a nondissipative medium with local interaction are called reciprocal elements if they do not change their sign for a time-frequency reversal; the elements which change sign are nonreciprocal elements.

An inspection of the matrix form (6.21) of the constitutive tensor shows that for a time reversal

$$\left.\begin{array}{l} \varepsilon_{lk} \to \varepsilon_{lk} \\ \chi_{lk} \to \chi_{lk} \\ \tilde{\gamma}_{lk} \to -\tilde{\gamma}_{lk} \\ \tilde{\gamma}^*_{lk} \to -\tilde{\gamma}^*_{lk} \end{array}\right\} \text{for } t \to -t , \tag{8.4}$$

because only the off-diagonal submatrices are associated with tensor elements with an odd number of time indices (any element with three zero indices vanishes identically due to (6.13)).

From the reality of the kernel $\varepsilon(t - \tau)$ (§ 6.1), we found that the real parts of the coefficients ε, χ and $\tilde{\gamma}$ are always even functions of frequency, while the imaginary parts are purely odd functions of frequency. These coefficients are functions of frequency only if one assumes local but noninstantaneous interaction. Hence an application of (8.3) gives

$$\left.\begin{array}{ll} \operatorname{Re}\varepsilon_{lk} \to \operatorname{Re}\varepsilon_{lk} & \operatorname{Im}\varepsilon_{lk} \to -\operatorname{Im}\varepsilon_{lk} \\ \operatorname{Re}\chi_{lk} \to \operatorname{Re}\chi_{lk} & \operatorname{Im}\chi_{lk} \to -\operatorname{Im}\chi_{lk} \\ \operatorname{Re}\tilde{\gamma}_{lk} \to \operatorname{Re}\tilde{\gamma}_{lk} & \operatorname{Im}\tilde{\gamma}_{lk} \to -\operatorname{Im}\tilde{\gamma}_{lk} \\ \operatorname{Re}\tilde{\gamma}^*_{lk} \to \operatorname{Re}\tilde{\gamma}^*_{lk} & \operatorname{Im}\tilde{\gamma}^*_{lk} \to -\operatorname{Im}\tilde{\gamma}_{lk} \end{array}\right\} \text{ for } \omega \to -\omega\,. \tag{8.5}$$

The simultaneous application of a time and frequency reversal then leads, according to (8.4) and (8.5) to the following table:

Reciprocal	*Nonreciprocal*
$\operatorname{Re}\varepsilon_{lk} \to \operatorname{Re}\varepsilon_{lk}$	$\operatorname{Im}\varepsilon_{lk} \to -\operatorname{Im}\varepsilon_{lk}$
$\operatorname{Re}\chi_{lk} \to \operatorname{Re}\chi_{lk}$	$\operatorname{Im}\chi_{lk} \to -\operatorname{Im}\chi_{lk}$
$\operatorname{Im}\tilde{\gamma}_{lk} \to \operatorname{Im}\tilde{\gamma}_{lk}$	$\operatorname{Re}\tilde{\gamma}_{lk} \to -\operatorname{Re}\tilde{\gamma}_{lk}$
$\operatorname{Im}\tilde{\gamma}^*_{lk} \to \operatorname{Im}\tilde{\gamma}^*_{lk}$	$\operatorname{Re}\tilde{\gamma}^*_{lk} \to -\operatorname{Re}\tilde{\gamma}_{lk}$
for $t \to -t$ and $\omega \to -\omega$	

(8.6)

We should add to this the information that ε_{lk} and χ_{lk} are Hermitian for a nondissipative medium. A constant contribution, independent of frequency, in the elements of ε_{lk}, χ_{lk} and $\tilde{\gamma}_{lk}$ should be regarded as an even (zero power) function of frequency; they occur only in the real parts. The imaginary elements, therefore, if different from zero, should be functions of frequency under all circumstances. One can express this in another way by saying that the imaginary elements are associated with effects which are essentially dispersive. We may illustrate these features in the following diagrams of the constitutive tensor:

Real components of $\chi^{\lambda\nu\sigma\kappa}$
(not necessarily dispersive)

If not isotropic, dielectric birefringence (reciprocal)	Fresnel-Fizeau effect in nonstationary media (nonreciprocal)	(8.7a)
Fresnel-Fizeau effect in nonstationary media (nonreciprocal)	If not isotropic, magnetic birefringence (reciprocal)	

Imaginary components of $\chi^{\lambda\nu\sigma\kappa}$
(essentially dispersive)

Dielectric Faraday effect (nonreciprocal)	Natural optical activity (reciprocal)	(8.7b)
Natural optical activity (reciprocal)	Magnetic Faraday effect (nonreciprocal)	

	Reciprocal	Nonreciprocal	
Not necessarily dispersive	Re ε Re χ	Re $\tilde{\gamma}$	(8.8)
Essentially dispersive	Im $\tilde{\gamma}$	Im ε Im χ	

The three-by-three submatrices of (6.21) are all tensors in the sense that they obey homogeneous transformation laws for spatial changes of the reference system. There is, however, a difference between the diagonal and off-diagonal submatrices with respect to inversions and mirror operations. This distinction is expressed by the fact that

$$\varepsilon_{lk} \quad \text{and} \quad \chi_{lk}$$

are tensors and

$$\tilde{\gamma}_{lk} \ (\text{and } \tilde{\gamma}^*_{lk})$$

are pseudo tensors, because $\tilde{\gamma}_{lk}$ relates two vectors of a different sort: a polar vector $\boldsymbol{E}$ with an axial vector $\boldsymbol{H}$ (or $\boldsymbol{D}$ with $\boldsymbol{B}$). The name tensor for a two index quantity frequently implies symmetry in the two indices. The elements $\tilde{\gamma}_{lk}$, however, are not symmetric in l and k; ε_{lk} and χ_{lk} have Hermitian symmetry only.

It was noted earlier that the elements $\tilde{\gamma}_{lk}$ for a stationary medium should be purely imaginary; they then represent the phenomenon of optical activity. One can now ask the question in the sense of the Neumann principle of crystal physics: What stationary media admit (imaginary) coefficients $\tilde{\gamma}_{lk}$ different from zero? One may follow a procedure indicated by SCHOUTEN [1951] by considering the decomposition:

$$\begin{aligned}&\text{A nonsymmetric pseudo tensor}\\ &\qquad = \text{a symmetric pseudo tensor} + \text{a vector}\,. \qquad (8.9)\end{aligned}$$

For the nonsymmetric tensor one has, in a similar way,

$$\begin{aligned}&\text{A nonsymmetric tensor} = \text{a symmetric tensor}\\ &\qquad + \text{a pseudo vector}. \qquad (8.10)\end{aligned}$$

Schouten uses the terminology "*W* tensor" for pseudo tensor and "affinor" for a nonsymmetric tensor. The above rules in Schouten's terminology are

$$W\,\text{affinor} = W\,\text{tensor} + \text{vector}\,, \qquad (8.9\text{a})$$

$$\text{affinor} = \text{tensor} + W\,\text{vector}\,. \qquad (8.10\text{a})$$

These decompositions (8.9) and (8.10) permit the application of the standard classification results of crystal physics for tensors and vectors. To obtain the classification for the pseudo affinor of the natural optical activity one applies (8.9a) thus arriving at the following set of matrices for the 32 crystal classes:

$$\begin{pmatrix} \tilde{\gamma}_{11} & \tilde{\gamma}_{12} & \tilde{\gamma}_{13} \\ \tilde{\gamma}_{21} & \tilde{\gamma}_{22} & \tilde{\gamma}_{23} \\ \tilde{\gamma}_{31} & \tilde{\gamma}_{32} & \tilde{\gamma}_{33} \end{pmatrix} 2 \quad \begin{pmatrix} 0 & 0 & \tilde{\gamma}_{13} \\ 0 & 0 & \tilde{\gamma}_{23} \\ \tilde{\gamma}_{31} & \tilde{\gamma}_{32} & 0 \end{pmatrix} 4$$

$$\begin{pmatrix} \tilde{\gamma}_{11} & \tilde{\gamma}_{12} & 0 \\ \tilde{\gamma}_{21} & \tilde{\gamma}_{22} & 0 \\ 0 & 0 & \tilde{\gamma}_{33} \end{pmatrix} 5 \quad \begin{pmatrix} \tilde{\gamma}_{11} & 0 & 0 \\ 0 & \tilde{\gamma}_{22} & 0 \\ 0 & 0 & \tilde{\gamma}_{33} \end{pmatrix} 7$$

$$\begin{pmatrix} 0 & \tilde{\gamma}_{12} & 0 \\ \tilde{\gamma}_{21} & 0 & 0 \\ 0 & 0 & 0 \end{pmatrix} 8 \quad \begin{pmatrix} \tilde{\gamma}_{11} & 0 & 0 \\ 0 & \tilde{\gamma}_{11} & 0 \\ 0 & 0 & \tilde{\gamma}_{33} \end{pmatrix} \text{10, 15, 22, media with rotational symmetry around the 3 axis}$$

$$\begin{pmatrix} 0 & \tilde{\gamma}_{12} & 0 \\ -\tilde{\gamma}_{12} & 0 & 0 \\ 0 & 0 & 0 \end{pmatrix} \text{11, 16, 23} \quad \begin{pmatrix} \tilde{\gamma}_{11} & \tilde{\gamma}_{12} & 0 \\ -\tilde{\gamma}_{12} & \tilde{\gamma}_{11} & 0 \\ 0 & 0 & \tilde{\gamma}_{33} \end{pmatrix} \text{13, 18, 25}$$

$$\begin{pmatrix} \tilde{\gamma}_{11} & 0 & 0 \\ 0 & -\tilde{\gamma}_{11} & 0 \\ 0 & 0 & 0 \end{pmatrix} 19 \quad \begin{pmatrix} \tilde{\gamma}_{11} & \tilde{\gamma}_{12} & 0 \\ \tilde{\gamma}_{12} & -\tilde{\gamma}_{11} & 0 \\ 0 & 0 & 0 \end{pmatrix} 20 \qquad (8.11)$$

$$\begin{pmatrix} \tilde{\gamma}_{11} & 0 & 0 \\ 0 & \tilde{\gamma}_{11} & 0 \\ 0 & 0 & \tilde{\gamma}_{11} \end{pmatrix} \text{29, 32} \quad \left\{ \text{rotationally symmetric media around arbitrary axes, no symmetry center.} \right.$$

Optical activity does not occur in the crystal classes 1, 3, 6, 9, 12, 14, 17, 21, 24, 26, 27, 28, 30, 31, nor in isotropic media (isotropy here defined by arbitrary rotational symmetry + symmetry center).

The numbers behind the matrices are the numbers characterizing the 32 classes according to VOIGT [1910]. The table shows how these numbers are associated with a generating set of symmetry operations defining each crystal class.

TABLE

1	C	9	C, z_3, x_2	17	C, z_4	25	z_6
2	—	10	z_3, x_2	18	z_4	26	z_3, x_2, E_z
3	C, z_2	11	z_3, E_x	19	S_z, x_2	27	$z_3\, E_z$
4	E_z	12	C, z_3	20	S_z	28	C, x_4, y_4
5	z_2	13	z_3	21	C, z_6, x_2	29	x_4, y_4
6	C, z_2, x_2	14	C, z_4, x_2	22	z_6, x_2	30	S_x, S_y
7	z_2, x_2	15	z_4, x_2	23	z_6, E_x	31	C, x_2, y_2, S
8	z_2, E_x	16	z_4, E_x	24	C, z_6	32	x_2, y_2, S

The definitions of the symmetry symbols are:

x_2 = x axis is twofold axis.
z_2 = z axis is twofold axis, etc.
x_3 = threefold axis, etc., for y and z.
x_4 = fourfold axis, etc., for y and z.
C = symmetry center.
S = cyclic permutation of axes.
E_x = reflection with respect to yz plane.
S_x = rotation around x over 90° + reflection with respect to yz plane, etc., for y and z.

The optical activity commonly discussed in the literature is the case characterized by the last matrix of (8.11). Optical activity shows itself there as a phenomenon independent of dielectric and magnetic birefringence. In all other cases, optical activity and birefringence occur simultaneously. There are, of course, directions in the crystals where the birefringence degenerates (e.g., the z axis of quartz, crystal class 10); then optical activity is again unperturbed by birefringence.

It should be kept in mind that the coefficients shown in (8.11) are the only ones that can exist, by virtue of crystal symmetry. There can be other physical reasons why some of these coefficients still vanish, although they could exist on basis of crystal symmetry.

The classification of the affinors ε_{lk} and χ_{lk} runs along similar lines as the classification of $\tilde{\gamma}_{lk}$. Instead of (8.9) one now uses the decomposition (8.10) (8.10a). In addition one has the information that both affinors ε_{lk} and χ_{lk} are Hermitian. The W vector, therefore, according to (8.10a), is given by a set of three purely imaginary components representing the Faraday effect. An inspection of the relevant sources of information about the application of the Neumann principle† shows that the W vector (e.g., the Faraday effect) may occur in media of the symmetry classes

$$1, 2, 3, 4, 5, 12, 13, 17, 18, 20, 24, 25, 27, 29, 32 \quad (8.12)$$

or in any medium which has rotational symmetry around one fixed axis.

The alternate name of "forced optical activity" for the Faraday effect is reminiscent of the fact that the Faraday effect does not commonly occur spontaneously without an external biasing field. Materials in a spontaneously magnetized state (e.g., ferrites), however, do exhibit Faraday rotation. Their original symmetry class should be regarded as being affected by the internal field. Please note that the biasing field itself must be a W vector. Therefore, magnetic fields or rotations (related by Larmor's theorem) may be considered as

† SCHOUTEN [1951], Chapter VII, Table 5.17.

generating agents of the Faraday effect. Furthermore it should be kept in mind that one deals here with a linearization of a basically nonlinear phenomenon, otherwise it would be contrary to our previously adopted rules to consider biasing fields as constituents of the constitutive tensor.

2. THE DIRECTIVE EFFECT AND THE FRESNEL-FIZEAU EFFECT

The directive effect is associated with the possible existence of a directive coupling between the permanent electric and magnetic dipoles in an electromagnetic medium. The directive effect as such should be well distinguished from the effect of natural optical activity. An optically active medium generates a magnetic dipole moment if it is subjected to a change of the electric field and conversely. Optical activity therefore is associated with the imaginary coefficients in the matrix $\tilde{\gamma}_{lk}$ as given in (6.21), but the directive effect, in contradistinction, should be associated with the real components in $\tilde{\gamma}_{lk}$ and would consequently lead to nonreciprocal phenomena.

DEBYE [1925] suggested the possible existence of a directive effect in gaseous media with molecules that were known to have a permanent electric dipole moment as well as a permanent magnetic dipole moment. Subsequent experiments carried out by HUBER [1926], however, led to completely negative results. To explain the outcome of this experiment it was assumed that Huber might have dealt with a balanced gas mixture of mirror symmetric molecules. Later work, however, showed that it seemed more probable that the directions of the two dipole moments in the molecule itself were uncorrelated. VAN VLECK [1932] pointed out that the mirror symmetrical configurations are really degenerate states of the molecular system, belonging to the same energy eigenvalues. The superposition principle of quantum mechanics then leads to the conclusion that any intermediate state is acceptable as a possible state of the molecule. Hence there should be no directional correlation between the two dipole moments. It is, of course, tacitly assumed that these considerations hold for stationary molecules. The same could still be true for a random motion of the molecules if one takes the time average, or for an assembly of randomly moving molecules if one takes the space average. We will assume that the aforementioned arguments are sufficient evidence to postulate the absence of a directive effect in all stationary matter. One then has adequate grounds to justify the restrictive relation (6.18). This relation in the form (6.18b) only claims that the trace of the pseudo tensor $\tilde{\gamma}_{lk}$ should vanish.

TELLEGEN [1948] brought up again this subject matter in a more recent investigation of nonreciprocal phenomena. He considered an assembly of electric-magnetic dipole twins, all of them lined up in the same fashion (either parallel or anti-parallel). Macroscopically, there is no reason to assume that these dipoles cannot be glued together, pairwise in an almost rigid manner. One then embeds the rigid dipole twins (randomly) in an ideally elastic binder and one seems to have obtained a medium which violates the relation (6.18).

It is not the place here to investigate in microphysical detail why nature does not provide what mankind seems to be able to make on a macroscopic scale. However, keeping in line with the formal scope of this text, we may investigate the wave propagation in a medium which violates the relation (6.18). Let us consider a medium with the constitutive tensor (6.21) in the form

$$\chi^{\lambda\nu\sigma\kappa} = \begin{array}{|c|c|} \hline -\varepsilon\delta_{lk} & \tilde{\gamma}\delta_{lk} \\ \hline \tilde{\gamma}\delta_{lk} & \dfrac{1}{\mu}\delta_{lk} \\ \hline \end{array} \tag{8.13}$$

The tensors ε_{lk}, χ_{lk} and the pseudo tensor $\tilde{\gamma}_{lk}$ all have the unit form, which means that the medium has arbitrary rotational symmetry, but no center of symmetry, because $\tilde{\gamma} \neq 0$.

Substitution of (8.13) into (6.48), with the help of (6.41), (6.42) and (6.43), yields the following explicit determinant equation for the phase velocities:

$$\left| \begin{pmatrix} 0 & 0 & 0 & 0 \\ 0 & \varepsilon & 0 & 0 \\ 0 & 0 & \varepsilon & 0 \\ 0 & 0 & 0 & \varepsilon \end{pmatrix} u^2 + \begin{pmatrix} 0 & -\varepsilon & 0 & 0 \\ -\varepsilon & 0 & 0 & 0 \\ 0 & 0 & 0 & 0 \\ 0 & 0 & 0 & 0 \end{pmatrix} u + \begin{pmatrix} \varepsilon & 0 & 0 & 0 \\ 0 & 0 & 0 & 0 \\ 0 & 0 & -1/\mu & 0 \\ 0 & 0 & 0 & -1/\mu \end{pmatrix} \right| = 0 . \tag{8.14}$$

The nontrivial roots of this equation are simply

$$u = \frac{\pm 1}{\sqrt{\varepsilon\mu}} \, \{\text{each root counted twice}\} ,^{\dagger} \tag{8.15}$$

thus leading to the conclusion that the coefficient $\tilde{\gamma}$ of the directive effect does not occur in (8.15), which means that it does not affect at all the propagation of electromagnetic waves. This result is not too surprising because we knew already that the condition (6.18) was associated with a so-called trivial Lagrangian which leads to an identically vanishing contribution in the Euler-Lagrangian derivative. We therefore assumed that the relations (6.18) and (6.19) are physically relevant restrictions for linear electromagnetic media. The covariance of (6.18) and (6.19) ascertained their validity for media in an arbitrary state of motion.

Summarizing, we seem to come to the conclusion that the only physically realistic directive effect occurs in nonstationary media in the form of the Fresnel-Fizeau effect. We may just

† The determinant (8.14) reduces to a two-by-two determinant for the nontrivial roots. Compare § 6.4.

briefly check whether our present formalism leads to the customary results that light waves are dragged along by the medium in the ratio $\alpha = (1 - 1/\varepsilon\mu)$.

The constitutive tensor for a translating isotropic medium has the form given by (6.23) in chapter 6, obtained by means of the infinitesimal Lorentz transformation. Please note that the configuration (6.23) satisfies (6.18). The substitution of (6.23) in (6.48), using again the matrices defined by (6.41), (6.42) and (6.43), yields the following determinant equation

$$\begin{vmatrix} \varepsilon & -\varepsilon u & \varepsilon\alpha v_2 & \varepsilon\alpha v_3 \\ -\varepsilon u & \varepsilon u^2 & -\varepsilon\alpha v_2 u & -\varepsilon\alpha v_3 u \\ \varepsilon\alpha v_2 & -\varepsilon\alpha v_2 u & (\varepsilon u^2 + 2\alpha\varepsilon v_1 u - 1/\mu) & 0 \\ \varepsilon\alpha v_3 & -\varepsilon\alpha v_3 u & 0 & (\varepsilon u^2 + 2\alpha\varepsilon v_1 u - 1/\mu) \end{vmatrix} = 0 \,. \tag{8.16}$$

The nontrivial roots of this equation (8.16) are given by the doubly counted roots of

$$\varepsilon u^2 + 2\alpha\varepsilon v_1 u - 1/\mu = 0 \,.$$

Discarding higher order terms ($\alpha v_1 \ll \sqrt{(1/\varepsilon\mu)}$) we find the familiar result

$$u_{1,2} = -\alpha v_1 \pm \sqrt{(1/\varepsilon\mu)} \quad \text{(counted twice)}\,. \tag{8.17}$$

3. THE FARADAY EFFECTS AND THE NATURAL OPTICAL ACTIVITY

Having dealt with the Fresnel-Fizeau effect in the previous paragraph, we may treat in a similar way as outlined in chapter 6 the Faraday effects and the natural optical activity. We will write down first the constitutive tensors in the form (6.21), then the corresponding determinant equations (6.47) and subsequently the nontrivial roots of these determinant equations. The formulae for the dielectric Faraday effect will be denoted with a label (a), the magnetic Faraday effect with a label (b), and the natural optical activity with a label (c).

The following forms of the constitutive tensors can be extracted from the literature

$$\left[\begin{array}{ccc|ccc} -\varepsilon_{11} & 0 & 0 & & & \\ 0 & -\varepsilon & -i\varepsilon_{23} & & \tilde{\gamma}_{lk}=0 & \\ 0 & i\varepsilon_{23} & -\varepsilon & & & \\ \hline & & & 1/\mu & 0 & 0 \\ & \tilde{\gamma}^{*}_{lk}=0 & & 0 & 1/\mu & 0 \\ & & & 0 & 0 & 1/\mu \end{array}\right] \tag{8.18a}$$

dielectric Faraday effect; magnetic field in the x direction (BORN [1932]).

$$\left[\begin{array}{ccc|ccc} -\varepsilon & 0 & 0 & & & \\ 0 & -\varepsilon & 0 & & \tilde{\gamma}_{lk}=0 & \\ 0 & 0 & -\varepsilon & & & \\ \hline & & & \chi_{11} & 0 & 0 \\ & \tilde{\gamma}^{*}_{lk}=0 & & 0 & \chi & i\chi_{23} \\ & & & 0 & -i\chi_{23} & \chi \end{array}\right] \tag{8.18b}$$

magnetic Faraday effect; magnetic field in the x direction (POLDER [1949]).

$$\left[\begin{array}{ccc|ccc} -\varepsilon & 0 & 0 & i\tilde{\gamma} & 0 & 0 \\ 0 & -\varepsilon & 0 & 0 & i\tilde{\gamma} & 0 \\ 0 & 0 & -\varepsilon & 0 & 0 & i\tilde{\gamma} \\ \hline -i\tilde{\gamma} & 0 & 0 & 1/\mu & 0 & 0 \\ 0 & -i\tilde{\gamma} & 0 & 0 & 1/\mu & 0 \\ 0 & 0 & -i\tilde{\gamma} & 0 & 0 & 1/\mu \end{array}\right] \tag{8.18c}$$

natural optical activity; medium with arbitrary rotational symmetry (EYRING, WALTER and KIMBALL [1944]).

The matrices (8.18) are being used to specify the matrices (6.41), (6.42) and (6.43), to form subsequently the determinant equations (6.47) corresponding to the cases a, b and c. They are

$$\begin{vmatrix} \varepsilon_{11} & -\varepsilon_{11}u & 0 & 0 \\ -\varepsilon_{11}u & \varepsilon_{11}u^2 & 0 & 0 \\ 0 & 0 & (\varepsilon u^2 - 1/\mu) & \mathrm{i}\varepsilon_{23}u^2 \\ 0 & 0 & -\mathrm{i}\varepsilon_{23}u^2 & (\varepsilon u^2 - 1/\mu) \end{vmatrix} = 0 \quad \text{dielectric Faraday effect} \tag{8.19a}$$

$$\begin{vmatrix} \varepsilon & -\varepsilon_{11}u & 0 & 0 \\ -\varepsilon_{11}u & \varepsilon_{11}u^2 & 0 & 0 \\ 0 & 0 & (\varepsilon u^2 - \chi) & \mathrm{i}\chi_{23} \\ 0 & 0 & -\mathrm{i}\chi_{23} & (\varepsilon u^2 - \chi) \end{vmatrix} = 0 \quad \text{magnetic Faraday effect} \tag{8.19b}$$

$$\begin{vmatrix} \varepsilon & -\varepsilon u & 0 & 0 \\ -\varepsilon u & \varepsilon u^2 & 0 & 0 \\ 0 & 0 & (\varepsilon u^2 - 1/\mu) & 2\mathrm{i}\tilde{\gamma}u \\ 0 & 0 & -2\mathrm{i}\tilde{\gamma}u & (\varepsilon u^2 - 1/\mu) \end{vmatrix} = 0 \quad \text{natural optical activity} \tag{8.19c}$$

The nontrivial roots of these equations for the phase velocity u are

$$u = \pm \sqrt{\left(\frac{1}{(\varepsilon \pm \varepsilon_{23})\mu}\right)} \quad \text{dielectric Faraday effect} \tag{8.20a}$$

$$u = \pm \sqrt{\left(\frac{\chi \pm \chi_{23}}{\varepsilon}\right)} \quad \text{magnetic Faraday effect} \tag{8.20b}$$

$$u = \pm \frac{\tilde{\gamma}}{\varepsilon} \pm \sqrt{\left(\frac{1}{\varepsilon\mu} + \frac{\tilde{\gamma}^2}{\varepsilon^2}\right)} \quad \text{natural optical activity} . \tag{8.20c}$$

Substituting the eigenvalues (8.20) (phase velocities u) in the corresponding determinants (8.19), one obtains the explicit form of the matrix of the homogeneous equation (6.40), corresponding to the three cases a, b and c and their eigenvalues. We know already that A_0 can be taken zero, an inspection of the determinants (8.19) then shows that for $u \neq 0$, A_1 should be taken zero. The nontrivial parts of the matrices are, therefore, the two-by-two submatrices in the lower right-hand corner. For all three cases a, b and c one finds only the following two matrices, each corresponding to two roots

$$\begin{pmatrix} 1 & \mathrm{i} \\ -\mathrm{i} & 1 \end{pmatrix}, \quad \begin{pmatrix} -1 & \mathrm{i} \\ -\mathrm{i} & -1 \end{pmatrix}. \tag{8.21}$$

The eigenvectors associated with these matrices (i.e., annulled by these matrices) are

$$\hat{A}_+ = (0, 0, 1, \mathrm{i}), \quad \hat{A}_- = (0, 0, 1, -\mathrm{i}). \tag{8.22}$$

The components $\hat{A}_2$ and $\hat{A}_3$ in (8.22) have mutual phase angles of 90 degrees and equal magnitude. Multiplying these vectors with the time exponential, one finds that a $\frac{1}{2}\pi$ phase angle then leads to a circular motion of the loci of the vectors A_+ and A_- in the yz plane as indicated below in the rectangular frames

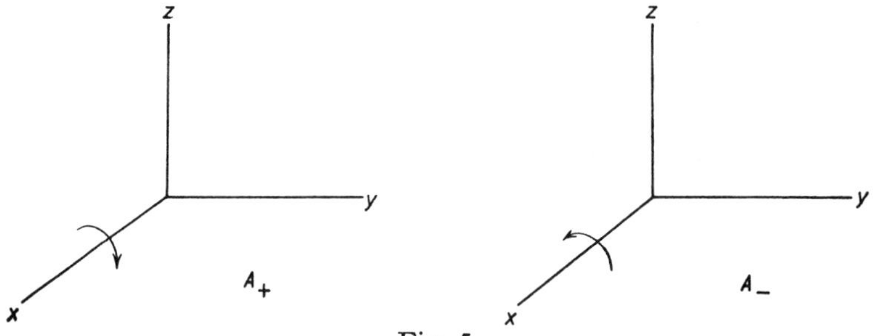

Fig. 5.

The eigenvalues are associated in the following way with the eigenvectors A_+ and A_-

$$A_+ \begin{cases} \pm\sqrt{\left(\dfrac{1}{(\varepsilon-\varepsilon_{23})\mu}\right)} & \text{dielectric Faraday effect, case (a)} \\[2ex] \pm\sqrt{\left(\dfrac{\chi+\chi_{23}}{\varepsilon}\right)} & \text{magnetic Faraday effect, case (b)} \\[2ex] \left\{\begin{matrix} +\dfrac{\tilde{\gamma}}{\varepsilon}+\sqrt{\left(\dfrac{1}{\varepsilon\mu}+\dfrac{\tilde{\gamma}^2}{\varepsilon^2}\right)} \\[2ex] +\dfrac{\tilde{\gamma}}{\varepsilon}-\sqrt{\left(\dfrac{1}{\varepsilon\mu}+\dfrac{\tilde{\gamma}^2}{\varepsilon^2}\right)} \end{matrix}\right\} & \text{natural optical activity, case (c)} \end{cases} \tag{8.23}$$

$$A_- \begin{cases} \pm\sqrt{\left(\dfrac{1}{(\varepsilon+\varepsilon_{23})\mu}\right)} & \text{dielectric Faraday effect, case (a)} \\[2ex] \pm\sqrt{\left(\dfrac{\chi-\chi_{23}}{\varepsilon}\right)} & \text{magnetic Faraday effect, case (b)} \\[2ex] \left\{\begin{matrix} -\dfrac{\tilde{\gamma}}{\varepsilon}+\sqrt{\left(\dfrac{1}{\varepsilon\mu}+\dfrac{\tilde{\gamma}^2}{\varepsilon^2}\right)} \\[2ex] -\dfrac{\tilde{\gamma}}{\varepsilon}-\sqrt{\left(\dfrac{1}{\varepsilon\mu}+\dfrac{\tilde{\gamma}^2}{\varepsilon^2}\right)} \end{matrix}\right\} & \text{natural optical activity, case (c)} \end{cases} \tag{8.24}$$

The root arrangements for the two eigenvectors give an idea about the differences between the Faraday rotation and the natural optical rotation. From (8.23) and (8.24) it is seen that, for the Faraday rotation, phase velocities of equal magnitude but opposite sign are associated either with A_+ or with A_-. Knowing that linearly polarized waves can be regarded as the superposition of two circularly polarized waves, this means that the Faraday effects (cases a and b) are characterized by a rotation of the plane of polarization independent of the sense of direction of propagation. Hence the screw sense of rotation and propagation sense is affected by a reversal of propagation.

In case (c) of the natural optical activity, however, phase velocities of equal magnitudes, but opposite signs, alternate between A_+ and A_-. Translated in terms of the linearly polarized waves, rotation and propagation sense are always associated according to the same screw sense. The screw sense characterizes the medium.

The differences, we just derived, between Faraday rotation and natural optical rotation are exactly the ones which enable the construction of nonreciprocal devices on the basis of the Faraday effect. Such a device is not possible on the basis of the natural optical rotation.

The characteristic features of the effects were all derived from (8.23) and (8.24) as relevant solutions of the generalized vector wave equation (6.28). The constitutive tensor occurring in this wave equation had to be specified accordingly. We found that the Faraday effect is spatially a pseudo vector effect, while the natural optical activity is a pseudo tensor effect, which may degenerate into a pseudo scalar effect. Only the pseudo scalar effect is associated with circular polarization for arbitrary directions of propagation. The pseudo vector and pseudo tensor each behave differently with respect to a time inversion due to the manner in which either is contained in the four-dimensional constitutive tensor.

An orderly and consistent application of time and space inversion brings out quite naturally all the essential features. It is not necessary to introduce the distinction between reciprocal and nonreciprocal effects as a physically required afterthought.

CHAPTER 9

MATTER-FREE SPACE WITH A GRAVITATIONAL FIELD

1. THE CONSTITUTIVE EQUATIONS OF MATTER-FREE SPACE

In this chapter we want to investigate an important form of the constitutive equations for matter-free space. First of all we may emphasize that there is really no observable experimental evidence that free space is nonlinear. We may therefore assume the equations to have the form already given by (6.12)

$$\mathfrak{G}^{\lambda\nu} = \tfrac{1}{2}\underset{0}{\chi}^{\lambda\nu\sigma\kappa}F_{\sigma\kappa}\,. \tag{9.1}$$

A zero label has been added to the kernel symbol of the constitutive tensor, conforming with common usage to denote free space relations.

In matter-free space we do not know any mechanisms that could be associated with a noninstantaneous interaction in the form of a delay of the response with respect to the applied fields. Nor do we know anything about a granular or discrete structure of matter-free space that might cause us to invoke the concept of nonlocal interaction, so that the displacement at a particular point would in addition depend on the field strength in its neighborhood. Hence, discarding the possibilities of noninstantaneous and nonlocal interactions, we find that the elements of the constitutive tensor for matter-free space must be purely real;

$$\underset{0}{\chi}^{\lambda\nu\kappa\sigma} \text{ is real}\,. \tag{9.2}$$

This implies, as discussed in chapter 6, that free space is absolutely nondispersive; a fact well supported by "light" velocity measurements in the visible and microwave ranges.

A comparison with diagram (8.7a) shows that a linear real relation of the type (9.1) can lead to two effects which are not commonly observed in free-space electromagnetics. They are birefringence or double refraction and a nonreciprocal effect of the Fresnel-Fizeau type.

It is hard to conceive of any mechanism which might bring about a dielectric or magnetic birefringence in free space. It should be noted that the absence of any polarization mechanism in free space is generally used as the standard argument to obliterate all field distinctions between $\boldsymbol{E}$, $\boldsymbol{D}$ and $\boldsymbol{H}$, $\boldsymbol{B}$ in matter-free space. We may now recall our earlier discussion in § 3.1 where we stressed the point that relations

$$\left.\begin{aligned} \boldsymbol{D} &= \boldsymbol{E} \\ \boldsymbol{B} &= \boldsymbol{H} \end{aligned}\right\} \quad \text{or} \quad \left\{\begin{aligned} \boldsymbol{D} &= \underset{0}{\varepsilon}\boldsymbol{E} \\ \boldsymbol{B} &= \underset{0}{\mu}\boldsymbol{H}, \end{aligned}\right. \tag{9.3}$$

really define a Lorentz invariant medium (provided one uses the proper units).

The relations (9.3) would be inadequate to describe propagation phenomena with respect to accelerated systems of coordinates, because they cannot be accommodated in the Lorentz group. Other physical agents, which might cause deviations from the uniform Lorentz invariant medium are gravitational fields.

Hence epitomizing the present discussion we conclude that the constitutive relations (9.3) are too confined (though adequate for many practical purposes); the relations (9.1) are too general to accommodate a fitting electromagnetic description of a (nonuniform) matter-free space. We must, therefore, look for a form of the constitutive equations that is intermediate between (9.1) and (9.3). We can impose as a physical requirement that this intermediate form is characterized by a generally invariant absence of birefringence, because there is no mechanism to produce it. The Fresnel-Fizeau type of asymmetry on the other hand must be regarded as an essential feature of a general nonuniform free space. Demanding the absence of both phenomena is not possible in the framework of general covariance; it would lead us back to the special case of which a Lorentz invariant medium characterized by (9.3) is an example.

An intermediate form between (9.1) and (9.3) can be obtained if we consider the case that $\underset{0}{\chi}^{\lambda\nu\sigma\kappa}$ has a degeneracy, so that it can be constructed from a tensor of lower valence.

The only field that comes to mind as a possible structural element of $\underset{0}{\chi}^{\lambda\nu\sigma\kappa}$ in matter-free space is the so-called "metrical" tensor of the space-time continuum in the sense of the theory of general relativity. It is known that the elements of this tensor $g_{\lambda\nu}$ or its inverse $g^{\lambda\nu}$ play a role in the interpretation of gravitational action. Gravitational action may be considered as a cause of nonuniformity in matter-free space. Thence, it should not be too far fetched to attempt to construct a $\underset{0}{\chi}^{\lambda\nu\sigma\kappa}$ from the elements of $g^{\lambda\nu}$ and $g_{\lambda\nu}$. A possible structure appears to be

$$\chi^{\lambda\nu\sigma\kappa} = \underset{0}{Y} g^{\frac{1}{2}}(g^{\lambda\sigma}g^{\nu\kappa} - g^{\lambda\kappa}g^{\nu\sigma}) \,, \tag{9.4}$$

with $\underset{0}{Y}$ a universal constant and g the absolute value of the determinant $| g_{\lambda\nu} |$. The Lagrangian density corresponding to the constitutive tensor (9.4) would be

$$\mathscr{L} = \tfrac{1}{8}\underset{0}{Y} g^{\frac{1}{2}} \{g^{\lambda\sigma}g^{\nu\kappa} - g^{\lambda\kappa}g^{\nu\sigma}\} F_{\lambda\nu}F_{\sigma\kappa} \,. \tag{9.5}$$

We will now show that the Lagrangian density (9.5) is a so-called admissible Lagrangian, because it satisfies the compatibility relation (5.39). Written in the form (5.46), the compatibility relation expresses that the energy momentum tensor is given by

$$\boxed{\mathfrak{T}_\mu^{\cdot\tau} = 2 g_{\mu\rho} \frac{\partial \mathscr{L}}{\partial g_{\tau\rho}}} \tag{9.6}$$

The Lagrangian density (9.5) is, however, expressed in the contravariant metrical components. It is therefore easier to use the equivalent relation

$$\boxed{\mathfrak{T}_\mu^{\ \tau} = -\, 2 g^{\tau\rho} \frac{\partial \mathscr{L}}{\partial g^{\mu\rho}}} \;\dagger \tag{9.6a}$$

Applying (9.6a) to the Lagrangian (9.5) we obtain

$$\mathfrak{T}_\mu^{\ \tau} = -\, F_{\lambda\nu} F_{\sigma\kappa} \frac{\underset{0}{Y} g^{\frac{1}{2}}}{4} g^{\tau\rho} \frac{\partial}{\partial g^{\mu\rho}} (g^{\lambda\sigma} g^{\nu\kappa} - g^{\lambda\kappa} g^{\nu\sigma})$$
$$-\, F_{\lambda\nu} F_{\sigma\kappa} \frac{\underset{0}{Y} g^{\frac{1}{2}}}{8} (g^{\lambda\sigma} g^{\nu\kappa} - g^{\lambda\kappa} g^{\nu\sigma}) g^{\tau\rho} \frac{\partial \ln g}{\partial g^{\mu\rho}} . \tag{9.7}$$

The second term in the right hand member of (9.7) is simply

† One can prove the equivalence of (9.6) and (9.6a) by the methods indicated in § 5.3, simply by assuming that $g^{\lambda\nu}$ instead of $g_{\lambda\nu}$ is the structural field. One can prove it directly too; it is then useful to recall the determinant relations

$$\mathrm{d} \ln g = g^{\lambda\nu} \mathrm{d} g_{\lambda\nu} = -\, g_{\lambda\nu} \mathrm{d} g^{\lambda\nu} \quad \text{with} \quad g = | g_{\lambda\nu} | .$$

equal to $\mathscr{L}\delta^{\tau}_{\mu}$, because, according to the determinant rule we just referred to in the footnote, we have that

$$-\frac{\partial \ln g}{\partial g^{\mu\rho}} g^{\tau\rho} = g_{\mu\rho}g^{\tau\rho} = \delta^{\tau}_{\mu} . \tag{9.8}$$

The reduction of the first term in the right hand member of (9.7) is slightly more cumbersome, but straightforward. Keeping in mind that

$$\mathfrak{G}^{\lambda\nu} = \tfrac{1}{2}\underset{0}{\chi}^{\lambda\nu\sigma\kappa}F_{\sigma\kappa} ,$$

we find that it reduces to

$$- F_{\mu\nu}\mathfrak{G}^{\tau\nu} . \tag{9.9}$$

Hence (9.7) becomes

$$\mathfrak{T}_{\mu}{}^{\tau} = \mathscr{L}\delta^{\tau}_{\mu} - F_{\mu\nu}\mathfrak{G}^{\tau\nu} , \tag{9.10}$$

the correct form of the energy momentum tensor as required by (9.6a) [see definition (4.55)].

Please note that it was essential for the proof to assume that $\underset{0}{Y}$ is a constant. If $\underset{0}{Y}$ were not a constant it would mean that we would have another structural field in addition to the metrical field $g^{\lambda\nu}$.

The relationship between the electromagnetic fields as given by the constitutive tensor (9.4) is very closely related to the relation originally postulated by Einstein†,

$$\mathfrak{G}^{\lambda\nu} = g^{\lambda\sigma}g^{\nu\kappa}F_{\sigma\kappa} . \tag{9.11}$$

In this form (9.11) it can not yet be identified with (9.4). However, interchanging the indices σ, κ in $F_{\sigma\kappa}$, the equation

† See LORENTZ, EINSTEIN, MINKOWSKI [1922] p. 117.

(9.11) becomes

$$\begin{aligned}\mathfrak{G}^{\lambda\nu} &= -\ g^{\lambda\sigma}g^{\nu\kappa}F_{\kappa\sigma} \\ &= -\ g^{\lambda\kappa}g^{\nu\sigma}F_{\sigma\kappa}\,.\end{aligned} \tag{9.12}$$

Adding (9.11) and (9.12) we find

$$\mathfrak{G}^{\lambda\nu} = \tfrac{1}{2}(g^{\lambda\sigma}g^{\nu\kappa} - g^{\lambda\kappa}g^{\nu\sigma})F_{\sigma\kappa}\,. \tag{9.13}$$

Einstein restricted his treatment to unimodular transformations ($g = 1$). Hence if we put in a factor $g^{\frac{1}{2}}$ to account for the density properties of the constitutive tensor and $\underset{0}{Y}$ to adjust the units we obtain exactly the form (9.4).

The reader may now well ask the question, whether it was really necessary to plow through a cumbersome compatibility proof, because a minor modification of the very acceptable relation (9.11) gives us the answers. Moreover, one may note that (9.11) degenerates into the customary field identification $\boldsymbol{E} = \boldsymbol{D}$ and $\boldsymbol{H} = \boldsymbol{B}$ on a Lorentz frame with light velocity $c = 1$.

We suppose, however, that it is necessary and useful to go through a more sophisticated compatibility argument, at least once, in order to assure oneself that generalizations of the kind as contained in (9.11) are justified. Moreover, one learns more about the assumptions and the conditions under which the generalizations are valid.

It is instructive to write the free-space constitutive tensor (9.4) in the matrix form (6.21)

$$\chi^{\lambda\nu\sigma\kappa} = \underset{0}{Y} g^{\frac{1}{2}} \left(\begin{array}{c|c} g^{00}g^{lk} - g^{0k}g^{0l} & g^{0s}g^{lk} - g^{0k}g^{ls} \\ \hline g^{s0}g^{kl} - g^{k0}g^{sl} & g^{lr}g^{ks} - g^{ls}g^{kr} \end{array} \right), \tag{9.14}$$

the zeros are time labels and $s, k, l, r = 1, 2, 3$ are space labels.

For an ideal (i.e., uniform) free space, we know that

$$\underset{0}{\chi}^{\lambda\nu\sigma\kappa} = \left(\begin{array}{c|c} -\underset{0}{\varepsilon} & 0 \\ \hline 0 & \underset{0}{\mu}^{-1} \end{array}\right), \tag{9.15}$$

the contravariant metric tensor $g^{\lambda\nu}$ becomes

$$g^{\lambda\nu} = \begin{pmatrix} \underset{0}{\varepsilon}\underset{0}{\mu} & 0 & 0 & 0 \\ 0 & -1 & 0 & 0 \\ 0 & 0 & -1 & 0 \\ 0 & 0 & 0 & -1 \end{pmatrix}, \tag{9.16}$$

and the determinant of the inverse of (9.16) has a square root of absolute value

$$g^{\frac{1}{2}} = (\underset{0}{\varepsilon}\underset{0}{\mu})^{-\frac{1}{2}}. \tag{9.17}$$

Substitution of (9.16) and (9.17) into (9.14) and comparison with (9.15) gives the following value for the constant $\underset{0}{Y}$

$$\underset{0}{Y} = \sqrt{\underset{0}{\varepsilon}/\underset{0}{\mu}}. \tag{9.18}$$

In the form (9.14) we may consider $\underset{0}{Y}$ as the gauge factor of the constitutive tensor for matter-free space. The dimension of $\underset{0}{Y}$ is according to (2.14) (chapter 2)

$$[\underset{0}{Y}] = [q^2\hbar^{-1}]. \tag{9.19}$$

Its numerical value in the MKS system, which is sometimes

called the admittance of free space, is

$$\underset{0}{Y} = \frac{1}{377} \text{ ohm}^{-1} . \tag{9.20}$$

The general structure of the free-space constitutive tensor (9.4) is not too dramatic from a physical viewpoint. There is, however, one factor which deserves some feature attention. If we look at the matrix form (9.14) we notice that there is a possibility for a nonreciprocal Fresnel-Fizeau-like asymmetry if the coefficients $g^{0s}(s = 1, 2, 3)$ of the metric do not vanish. The question arises whether there are configurations of gravitational fields or accelerated frames so that $g^{0s} \neq 0$.

There is an algorithm to compute the coefficients $g^{\lambda\nu}$ corresponding to a given physical situation, if we assume the general validity of the gravitational field equations of general relativity. One then finds that the elements g^{0s} are still zero for the Schwarzschild† solution, which corresponds to a central symmetric gravitational source. The situation begins to differ if one introduces time dependent gravitational fields. Please note, however, that the space-time anisotropy of the Fresnel-Fizeau effect is then, as a rule, coupled to a space-time nonuniformity, because the elements $g^{\lambda\nu}$ depend on the coordinates and on the time.

2. THE RIEMANN TENSOR AND THE CONSTITUTIVE TENSOR

In the previous pages we established a relation between the constitutive tensor of free-space and the metrical tensor of the space-time manifold. There is another example of a similar tensor which can be derived from the metric of a manifold, which is frequently discussed in the literature on differential geometry. This tensor is known as the Riemann or Riemann-Christoffel tensor. It depends in general directly on the elements

† See for instance TOLMAN [1934].

of the metric tensor, as well as on their first and second order derivatives. The Riemann-Christoffel tensor is what is known as a differential concomitant of the metric tensor.

The interpretation of the Riemann tensor is associated with the possibility of having linear frames of reference throughout a finite region of a manifold. A linear frame in this connection is a frame that is characterized by elements $g_{\lambda\nu}$ which are constants. If the Riemann tensor vanishes then it is possible to have a linear frame of reference. It is not possible to have a linear frame of reference if the Riemann tensor does not vanish.

Hence, a Riemann tensor of a metric whose elements are constants should be identically zero. It should still be zero after a transition to curvilinear coordinates, resulting in elements $g_{\lambda\nu}$ which are now functions of the coordinates; otherwise it would violate the homogeneous transformation law of tensors. A useful exercise to familiarize oneself with this complicated analytical criterion is to write out the Riemann tensor of these nonconstant elements $g_{\lambda\nu}$ in order to check that it still reduces to zero.

In other words, given an arbitrary covariant (or contravariant) tensor field of valence two, with elements depending on the coordinates, the Riemann tensor of this tensor provides a necessary and sufficient condition whether the original tensor of valence two can be transformed into a set of constant elements in a finite region of the manifold.

The present wording shows that it is by no means essential that the tensor of valence two represents a metric of a geometrical manifold. In point of fact, de Saint Vénant obtained the same criterion, discussing the deformation tensor of elasticity. An elastic body characterized by a deformation tensor which has a nonvanishing Riemann concomitant can never be "molded" into a state so that its deformation is constant throughout the body.

The analytical properties of the Riemann tensor have been

extensively discussed in the mathematical literature. It turns out to be a tensor of valence four, which, by means of contraction with the original tensor $g_{\lambda\nu}$ or its inverse $g^{\lambda\nu}$, can be written in an all covariant (or contravariant) form. Taking the all-covariant form, the Riemann tensor satisfies the following relations†

$$R_{\lambda\nu\sigma\kappa} = -R_{\nu\lambda\sigma\kappa} \tag{9.21}$$

$$R_{\lambda\nu\sigma\kappa} = -R_{\lambda\nu\kappa\sigma} \tag{9.22}$$

$$R_{\lambda\nu\sigma\kappa} = R_{\sigma\kappa\lambda\nu} \tag{9.23}$$

$$R_{[\lambda\nu\sigma\kappa]} = 0 \tag{9.24}$$

$$\nabla_{[\mu}R_{\lambda\nu]\sigma\kappa} = 0\,. \tag{9.25}$$

The last of these relations, the equation (9.25), is known as the Bianchi identity. The symbol ∇ denotes the covariant derivative. The Bianchi identity, incidentally, is of key importance in the formulation of the gravitational field equations of general relativity.

Now let us compare the relations (9.21), (9.22) , . . . , (9.25), which are characteristic for the Riemann tensor, with the relations we obtained for the constitutive tensor of an arbitrary linear electromagnetic medium, with instantaneous and local interaction between the fields. The relevant expressions were obtained in chapter 6, they are (6.13), (6.14), (6.18) and (6.19). We may repeat them here for the convenience of comparison,

$$\chi^{\lambda\nu\sigma\kappa} = -\chi^{\nu\lambda\sigma\kappa} \tag{9.26}$$

$$\chi^{\lambda\nu\sigma\kappa} = -\chi^{\lambda\nu\kappa\sigma} \tag{9.27}$$

$$\chi^{\lambda\nu\sigma\kappa} = \chi^{\sigma\kappa\lambda\nu} \tag{9.28}$$

† Compare SCHOUTEN [1951] p. 123.

$$\chi^{[\lambda\nu\sigma\kappa]} = 0 \qquad (9.29)$$

$$\partial_\nu \chi^{[\lambda\nu\sigma\kappa]} = 0 \,. \qquad (9.30)$$

The first four relations (9.26) . . . (9.29) are pairwise identical with the first four relations (9.21) . . . . (9.24). Consequently, the number of independent tensor components is the same. From the matrix form (6.21) one can easily derive that this number is $6 + 6 + 8 = 20$, because both tensors occur in a manifold of four dimensions. There are only eight independent components of $\tilde{\gamma}_{lk}$, because the $\tilde{\gamma}$ matrix has a vanishing trace [see (6.18a)], the numbers $6 + 6$ denote the number of possible dielectric and magnetic components.

A real difference occurs if we compare the Bianchi identity (9.25) with the identity (9.30) for the constitutive tensor. The Bianchi identity turns out to be a much more powerful restriction than (9.30).

We may now show that the identity (9.30) is in general contained in the Bianchi identity (9.25). To do this, we write out the alternation in (9.25),

$$\nabla_\mu R_{\lambda\nu\sigma\kappa} + \nabla_\nu R_{\mu\lambda\sigma\kappa} + \nabla_\lambda R_{\nu\mu\sigma\kappa} = 0 \,, \qquad (9.31)$$

next we multiply and contract with $g^{\mu\kappa}$, using the fact that the metric is constant for covariant differentiation,

$$\nabla_\mu R_{\lambda\nu\sigma}^{\cdot\cdot\cdot\mu} + \nabla_\nu R_{\lambda\sigma} - \nabla_\lambda R_{\nu\sigma} = 0 \,. \qquad (9.32)$$

Alternation over $\lambda\nu\sigma$ gives

$$\nabla_\mu R_{[\lambda\nu\sigma]}^{\cdot\cdot\cdot\mu} = 0 \,, \qquad (9.33)$$

because the contracted Riemann tensor $R_{\lambda\sigma}$, known as the Ricci tensor, cancels out, in view of the fact that it is symmetric in λ, σ. Raising the indices $\lambda\nu\sigma$ in (9.33) and adding the factor $g^{\frac{1}{2}}$ to provide the density properties gives the desired relation

$$\partial_\mu g^{\frac{1}{2}} R^{[\lambda\nu\sigma\mu]} = 0\,, \tag{9.34}$$

which is "isomeric" with (9.30). Please note that the index μ may be included in the alternation, because it alternates already with σ (see (9.22)). The covariant derivative degenerates into an ordinary derivative because the divergence of an alternating covariant density of weight $+1$ is a naturally invariant expression (§ 3.2).

Thus far, we have compared the mathematical properties of the Riemann tensor of a metric with the constitutive tensor of a rather arbitrary electromagnetic medium. We found that the Riemann tensor satisfies all the requirements of a constitutive tensor if we write it as a density of weight $+1$ in the contravariant form. We found also that it is not necessarily true that a constitutive tensor meets all conditions for a Riemann tensor, because (9.25) entails (9.30) but not the other way around.

We may now proceed to compare the more special constitutive tensor of matter-free space with the Riemann tensor associated with the metric of matter-free space. For the constitutive tensor we obtained the expression (9.4)

$$\underset{0}{\chi}^{\lambda\nu\sigma\kappa} = \underset{0}{Y} g^{\frac{1}{2}}(g^{\lambda\sigma}g^{\nu\kappa} - g^{\lambda\kappa}g^{\nu\sigma})\,. \tag{9.35}$$

An explicit expression of the Riemann tensor in terms of the metric may be extracted from the literature†. In the all-covariant form it is

$$R_{\lambda\nu\sigma\kappa} = g_{\kappa\tau}\{2\partial_{[\lambda}\Gamma^{\tau}_{\nu]\sigma} + 2\Gamma^{\tau}_{\mu[\lambda}\Gamma^{\mu}_{\nu]\sigma}\}\,. \tag{9.36}$$

† SCHOUTEN [1951].

The symbols $\Gamma^{\tau}_{\nu\sigma}$ represent the Christoffel expression which are related to the metric tensor according to (1.15).

At first sight there appears to be little chance that the simple algebraic concomitant (9.35) might be related to the complicated differential concomitant (9.36) of the metric. There is, however, an old theorem of differential geometry, known as Schur's theorem (LEVI CIVITA [1926]), which claims that there are manifolds in which (9.36) reduces indeed to a form equivalent to (9.35). The manifolds for which this is true have a Riemann scalar

$$R = R_{\lambda\nu\sigma\kappa} g^{\lambda\sigma} g^{\nu\kappa} , \tag{9.37}$$

which is a constant. The Riemann tensor is then given by the expression

$$R_{\lambda\nu\sigma\kappa} = \frac{R}{n(n-1)} (g_{\lambda\sigma} g_{\nu\kappa} - g_{\lambda\kappa} g_{\nu\sigma}) , \tag{9.38}$$

in which n is the number of dimensions of the manifold. The form (9.38), of course, leads to the contravariant form, with $n = 4$,

$$R^{\lambda\nu\sigma\kappa} = \tfrac{1}{12} R (g^{\lambda\sigma} g^{\nu\kappa} - g^{\lambda\kappa} g^{\nu\sigma}) . \tag{9.39}$$

Multiplication by $g^{\frac{1}{2}}$ produces the form (9.35) of the constitutive tensor.

Conversely, according to the Schur theorem, if the Riemann tensor has the form (9.38) then it can be proven that R is a constant.

It should be understood that the similarity between the constitutive tensor and the Riemann tensor of a manifold with constant Riemann scalar is only a statement of fact. It is by no means a physical proof that the matter-free domains of space-time should be regarded as having a constant Riemann scalar. It may be of some interest, however, to mention that

Einstein in 1919 discussed the possibility of constant "curvature" properties of matter-free space in connection with the structure of charged elementary particles. These arguments have now lost some impact and do not apply to the case of the later discovered neutral particles. But still, his considerations, entirely different from the arguments presented here, are up to a point independent of the application to elementary particles. After treating the conservation relations in the last paragraph of this chapter, we will have an opportunity to reconsider Einstein's very elegant and simple discussion.

In conclusion we want to emphasize that if (9.39) can be physically associated with (9.35) that then the dimensionless factor $\frac{1}{12}K$ (chapter 2) still requires a constant multiplier of dimension $[q^2\hbar^{-1}]$.

3. THE "EIKONAL" EQUATION AND THE NULL GEODESICS

It is a rather well known fact that the behavior of light, in the so-called geometrical optical approximation, has much similarity to the behavior of particles following well determined trajectories. Newton's emission theory of light and even the term "light rays" may be regarded as direct substantiations of this classical observation. We know now that the trajectory aspect of light loses its significance as soon as the medium has irregularities in it of the order of magnitude of the wave length of light. The body then becomes opaque. Hamilton was one of the first scientists who studied the limits of complementarity of the so-called wave and particle aspect of light. Following his procedure we may investigate the geometrical optical approximations of a solution of the wave equation (6.28). This wave equation, which has been discussed in chapter 6 and following, gave us an opportunity to formulate, in a covariant way, propagation problems in nonuniform media. At the same time it rendered us useful services for a study of propagation features in nonreciprocal and anisotropic media.

For the present purpose we may consider the propagation aspects in matter-free space, using the constitutive tensor (9.35), (9.4), as discussed in the two previous sections of this chapter. Considering this constitutive tensor, one expects the trajectories in matter-free space to depend on the space-time metric. We therefore have here an interesting opportunity to check whether the outcome of this operation conforms with the general relativistic postulate, according to which light rays should travel along null geodesics.

Writing down the wave equation (6.28) for a charge and current free space ($\mathfrak{c}^{\lambda} = 0$) we have

$$\partial_{\nu} \underset{0}{\chi}^{\lambda\nu\sigma\kappa} \partial_{\sigma} A_{\kappa} = 0 , \tag{9.40}$$

in which $\underset{0}{\chi}^{\lambda\nu\sigma\kappa}$ is given by (9.4). The next step is that we try to find a solution of (9.40) in accordance with the geometrical optical approximation. We may consider the trial solution

$$A_{\kappa} = a_{\kappa}\, \mathrm{e}^{\mathrm{i}S} \tag{9.41}$$

For a plane wave solution like (6.33) we had that a_{κ} is constant and S a linear function of the coordinates. We may now anticipate a deviation from plane wave behavior by assuming that a_{κ} is not constant any more but weakly dependent on the coordinates, whereas S may deviate slightly from linearity. In the form of an inequality this can be expressed by

$$\left| \frac{\partial_{\nu} a_{\kappa}}{a_{\kappa}} \right| \ll \partial_{\nu} S . \tag{9.42}$$

A specification of an "almost plane wave solution" (9.41), in the sense of (9.42), is justified only if the time and space-like irregularities of the medium are small with respect to the time-space periodicity of the wave. This can be expressed by the inequality

$$\left| \frac{\partial_\tau \underset{0}{\chi}^{\lambda\nu\sigma\kappa}}{\underset{0}{\chi}^{\lambda\nu\sigma\kappa}} \right| \ll \partial_\tau S \,. \tag{9.43}$$

A substitution of (9.41) in (9.40), taking into account the inequalities (9.42) and (9.43), yields as the predominant term

$$a_\kappa \chi^{\lambda\nu\sigma\kappa} (\partial_\nu S)\,(\partial_\sigma S) = 0 \,. \tag{9.44}$$

We may now introduce the expression (9.4) for the constitutive tensor into (9.44),

$$a_\kappa g^{\lambda\sigma} g^{\nu\kappa} (\partial_\nu S)\,(\partial_\sigma S) - a_\kappa g^{\lambda\kappa} g^{\nu\sigma} (\partial_\nu S)\,(\partial_\sigma S) = 0 \,. \tag{9.45}$$

The first term in (9.45) contains a factor

$$a_\kappa g^{\nu\kappa} \partial_\nu S \,,$$

which can be considered as a negligible entity, because it expresses the Lorentz condition (6.34).

Hence only the second term in (9.45) is of any consequence. For finite $a^\lambda = g^{\lambda\kappa} a_\kappa$, we therefore end up with the equation

$$\boxed{g^{\nu\sigma} (\partial_\nu S)\,(\partial_\sigma S) = 0} \tag{9.46}$$

The equation (9.46) is sometimes called the "eikonal" equation, because an equation of this type plays a role in the analysis of images in geometric optical applications.

For the present purpose we consider (9.46) as a Hamilton-Jacobi equation for the trajectories of light rays in matter-free space. This invites us to apply the Hamilton-Jacobi formalism of analytical dynamics.

We may now introduce the common abbreviations

$$H = \tfrac{1}{2}g^{\lambda\nu}k_\lambda k_\nu = 0\,, \tag{9.47}$$

with

$$k_\nu = \partial_\nu S\,. \tag{9.48}$$

The equations for the trajectories, also known as Hamilton's canonical equations then become

$$\boxed{\begin{aligned} \dot{x}^\lambda &= \frac{\partial H}{\partial k_\lambda} \\ \dot{k}_\nu &= -\frac{\partial H}{\partial x^\nu} \end{aligned}} \tag{9.49}$$

The dots over x^λ and k_ν denote a differentiation with respect to a scalar parameter, say u, along the trajectories.

The equations (9.49), in a mathematical sense, are the equations of the so-called characteristics associated with the nonlinear partial differential equation (9.47).

The system of eight first order ordinary differential equations can be combined into four second order equations of the type

$$\boxed{\frac{\mathrm{d}}{\mathrm{d}u}\frac{\partial L}{\partial \dot{x}^\nu} - \frac{\partial L}{\partial x} = 0} \tag{9.50}$$

with

$$L = k_\nu \dot{x}^\nu - H = \tfrac{1}{2}g_{\lambda\nu}\dot{x}^\lambda\dot{x}^\nu = 0\,. \tag{9.51}$$

The equation (9.50) is an equation of a well known type. It expresses the condition for which the integral of the line element,

$$\int \mathrm{d}s = \int \sqrt{(g_{\lambda\nu}\,\mathrm{d}x^{\lambda}\,\mathrm{d}x^{\nu})}\,, \tag{9.52}$$

should have an extreme value. Trajectories satisfying the condition

$$\delta \int \mathrm{d}s = 0\,, \tag{9.52a}$$

are called geodesics. We know already, according to (9.51), that $\mathrm{d}s = 0$. Hence the trajectories determined by (9.50) are indeed the null geodesics of matter-free space.

An equivalent form of (9.50) is

$$\boxed{\ddot{x}^{\nu} + \Gamma^{\nu}_{\lambda\kappa}\dot{x}^{\lambda}\dot{x}^{\nu} = 0} \tag{9.53}$$

(BRILLOUIN [1949], ch. 8) with $\Gamma^{\nu}_{\lambda\kappa}$ the Christoffel symbol defined according to (1.15).

An important feature, in making the transition from wave equation (9.40) to the trajectory equation (9.53), is the fact that one does not invoke the equivalence principle nor any other specific general relativistic postulates, except the principle of general covariance. In addition it should be noted that it is in general not possible to determine unambiguously a wave equation associated with the trajectory equation (9.53). More than one wave equation will lead to the geodesic (9.53). The reader is invited to check with the literature (VON LAUE [1920]; EDDINGTON [1923]; WHITTAKER [1928]; TOLMAN [1934], p. 267) to get an impression of the many different proposed forms of the vector wave equation for matter-free space. In making the transition from the Lorentz invariant d'Alembertian to a generally covariant vector wave equation, one has a case in point where one should be cautious to apply the recipe: ordinary differential → covariant differential. This procedure is in general less dangerous in a manifold with vanishing Riemann tensor; then covariant derivatives commute like ordinary derivatives.

4. ENERGY-MOMENTUM RELATIONS FOR MATTER-FREE SPACE

In § 4.3 we derived the quite general equation (4.49) for the energy momentum balance in an arbitrary conducting medium with local and instantaneous interaction between the fields. We may rewrite this equation here in the simple form

$$\boxed{\partial_\nu \mathfrak{T}_\lambda{}^\nu - \partial_{(\lambda)} \underset{0}{\mathscr{L}} = F_{\lambda\nu} \mathfrak{c}^\nu} \tag{9.54}$$

The tensor density $\mathfrak{T}_\lambda{}^\nu$ of energy and momentum in the relation (9.54) is defined by the expression

$$\mathfrak{T}_\lambda{}^\nu = \underset{0}{\mathscr{L}} \delta_\lambda^\nu - \mathfrak{G}^{\nu\sigma} F_{\lambda\sigma} \,. \tag{9.55}$$

It was shown that the energy momentum balance (9.54) should be regarded as a modified and extended version of the customary relation (4.44). The relation (4.44) was found to be valid only for linear and uniform media.

The new term $\partial_{(\lambda)} \underset{0}{\mathscr{L}}$, in equation (9.54), was denoted by the so-called explicit derivative of that part of the Lagrangian density ($\underset{0}{\mathscr{L}}$) which depended only on the derivatives of the potentials. The explicit derivative $\partial_{(\lambda)}$ was specified as an operation applying to the structural field of $\underset{0}{\mathscr{L}}$ only, implying that the potentials and their derivatives should be regarded as constants with respect to this differentiation.

The physical interpretation of the new term $\partial_{(\lambda)} \underset{0}{\mathscr{L}}$ was found to be associated with the force density on the irregularities in a nonuniform medium. It was shown, for instance, that the term $\partial_{(\lambda)} \underset{0}{\mathscr{L}}$ just balanced the radiation forces on the interface of two different but nonconducting media.

We may add to this recapitulation, that the expression (9.54) has been reproduced in chapter 5 [see formula (5.45)] by a different method, for the case of a nonconducting medium. Furthermore, we may mention that the specific form (9.55) of the energy momentum tensor occurred quite naturally in the compatibility relations (5.39), (5.46) for the Lagrangian density.

An application of the equation (9.54) to the case of matter-free space suggests an interpretation of the term $\partial_{(\lambda)} \underset{0}{\mathscr{L}}$, which should be associated with the nonuniformity of free space. The nonuniformity being due to the presence of gravitational fields one expects now $\partial_{(\lambda)} \underset{0}{\mathscr{L}}$ to play a role in the interaction of electromagnetic and gravitational forces. We may therefore compare (9.54) with the customary expression of general relativity in which the energy momentum balance is expressed by the covariant divergence

$$\nabla_\nu \mathfrak{T}_\lambda{}^\nu = 0 . \tag{9.56}$$

Writing out the covariant derivative one obtains

$$\nabla_\nu \mathfrak{T}_\lambda{}^\nu = \partial_\nu \mathfrak{T}_\lambda{}^\nu - \mathfrak{T}_\nu{}^\kappa \Gamma^\nu_{\kappa\lambda} = 0 . \tag{9.56a}$$

The right-hand term in (9.54) vanishes, because $\mathfrak{c}^\nu = 0$ in matter-free space. Hence comparing the expressions (9.54) and (9.56a) we may consider whether or not the following relation holds,

$$\boxed{\partial_{(\lambda)} \underset{0}{\mathscr{L}} = \mathfrak{T}_\nu{}^\kappa \Gamma^\nu_{\kappa\lambda}} \tag{9.57}$$

The proof that (9.57) is indeed true is actually quite simple if we take advantage of the fact that

$$\nabla_{(\lambda)} \underset{0}{\mathscr{L}} = 0\,, \tag{9.58}$$

because the metric which is the structural field in $\underset{0}{\mathscr{L}}$ is a covariant constant. Writing out the (explicit) covariant derivative of

$$\underset{0}{\mathscr{L}} = \tfrac{1}{2}\underset{0}{\chi}^{\tau\nu\sigma\kappa} F_{\tau\nu} F_{\sigma\kappa}\,, \tag{9.59}$$

we get

$$\begin{aligned}
\nabla_{(\lambda)} \underset{0}{\mathscr{L}} = \partial_{(\lambda)} \underset{0}{\mathscr{L}} & \\
&\left.\begin{aligned}
&+ \tfrac{1}{2} F_{\tau\nu} F_{\sigma\kappa} \underset{0}{\chi}^{\mu\nu\sigma\kappa} \Gamma^{\tau}_{\lambda\mu} \\
&+ \tfrac{1}{2} F_{\tau\nu} F_{\sigma\kappa} \underset{0}{\chi}^{\tau\mu\sigma\kappa} \Gamma^{\mu}_{\lambda\mu} \\
&+ \tfrac{1}{2} F_{\tau\nu} F_{\sigma\kappa} \underset{0}{\chi}^{\tau\nu\mu\kappa} \Gamma^{\lambda}_{\lambda\mu} \\
&+ \tfrac{1}{2} F_{\tau\nu} F_{\sigma\kappa} \underset{0}{\chi}^{\tau\nu\sigma\mu} \Gamma^{\kappa}_{\lambda\mu}
\end{aligned}\right\} \to \mathfrak{G}^{\mu\nu} F_{\tau\nu} \Gamma^{\tau}_{\lambda\mu} \\
&- \tfrac{1}{2} F_{\tau\nu} F_{\sigma\kappa} \underset{0}{\chi}^{\tau\nu\sigma\kappa} \Gamma^{\mu}_{\lambda\mu} \;\} \to - \underset{0}{\mathscr{L}} \Gamma^{\mu}_{\lambda\mu}\,.
\end{aligned} \tag{9.60}$$

The reduction of the terms in (9.60), as indicated by the arrows, can be performed in a straightforward manner by using the skew symmetry of the field $F_{\lambda\nu}$ and the relations (6.13) of the constitutive tensor.

Applying (9.58) to (9.60) gives the expression

$$\partial_{(\lambda)} \underset{0}{\mathscr{L}} = (\delta^{\mu}_{\tau} \underset{0}{\mathscr{L}} - \mathfrak{G}^{\mu\nu} F_{\tau\nu}) \Gamma^{\tau}_{\lambda\mu}\,, \tag{9.61}$$

which is equivalent to (9.57), because of the definition equation (9.55).

As a next point of discussion we may consider Einstein's argument for the constant "curvature" properties of matter-free space (LORENTZ, EINSTEIN and MINKOWSKI [1922] p. 140). Einstein remarked that the trace of the energy momentum tensor in electromagnetics vanishes,

$$\mathfrak{T}_{\lambda}{}^{\lambda} = 0\,. \tag{9.62}$$

One can easily confirm the relation (9.62) by means of the definition equation (9.55). We may add to this the observation that the vanishing trace characterizes a linear medium.

The energy momentum tensor, according to the principle of equivalence, should be associated with the Riemann tensor. The only concomitant of the Riemann tensor that could satisfy the requirement (9.62) is

$$(R_{\lambda}{}^{\nu} - \tfrac{1}{4}\delta_{\lambda}^{\nu}R)\,, \tag{9.63}$$

in which $R_{\lambda}{}^{\nu}$ is the contracted Riemann tensor (Ricci tensor) and R the Riemann scalar defined by (9.37).

The gravitational field equations for matter-free space with electromagnetic radiation as the source field would then be

$$g^{\frac{1}{2}}(R_{\lambda}{}^{\nu} - \tfrac{1}{4}\delta_{\lambda}^{\nu}R) = -\,(\kappa)\mathfrak{T}_{\lambda}{}^{\nu}\,, \tag{9.64}$$

with (κ) the gravitational constant of general relativity, which has the dimension $[\hbar^{-1}]$, but not the numerical value $\hbar^{-1}$.

Taking the covariant divergence of (9.64), one finds, because of (9.56), that

$$\nabla_{\nu}(R_{\lambda}{}^{\nu} - \tfrac{1}{4}\delta_{\lambda}^{\nu}R) = 0\,. \tag{9.65}$$

From the Bianchi identity we obtained the generally valid

equation (9.32). Contraction of (9.32) with $g^{\nu\sigma}$, however, yields

$$\nabla_\nu (R_\lambda{}^\nu - \tfrac{1}{2}\delta^\nu_\lambda R) = 0\,. \tag{9.66}$$

The expression (9.66) is valid in any Riemannian manifold. Hence the equations (9.65) and (9.66) can be compatible only if

$$\boxed{\partial_\lambda R = 0} \tag{9.67}$$

The conclusion contained in equation (9.67) may still not be considered as a proof that it is physically meaningful to regard the space-time manifold associated with matter-free space (with radiation as a source of the gravitational field) as a manifold with a constant Riemann scalar. A real physical proof (or rather support) can only be obtained from the crucial experiments of general relativity or modifications and extensions of those. In this connection it may be of some interest to mention that the conclusion that light rays should travel along null geodesics (9.53) may be regarded as being contained in (9.56). Suppose one considers a "pencil" of photons. The free-space medium is linear, hence the photons do not interact and their number is conserved. The energy momentum tensor for the pencil of photons can then be represented, according to (4.61) by

$$\mathfrak{T}_\lambda{}^\nu = \hbar k_\lambda \mathfrak{N}^\nu\,, \quad \text{with} \quad \partial_\nu \mathfrak{N}^\nu = 0\,. \tag{9.68}$$

Taking the covariant divergence of (9.68) one obtains according to (9.56)

$$\mathfrak{N}^\nu \nabla_\nu k_\lambda = 0\,. \tag{9.69}$$

The vector density $\mathfrak{N}^\nu$ of the "photon stream" can be written in the form

$$\mathfrak{N}^\nu = \underset{0}{\mathfrak{N}} \frac{dx^\nu}{du}\,. \tag{9.70}$$

Substitution of (9.70) into (9.69) yields the conclusion that the covariant differential of k_λ vanishes. This statement is equivalent with (9.53) if one considers the relations (9.49) and (9.47), according to which $\dot{x}^\lambda = g^{\lambda\nu}k_\nu$.

REFERENCES

BÔCHER, M., [1907], Introduction to Higher Algebra (MacMillan, New York).

BODE, H. W., [1945], Network Analysis and Feedback Amplifier Design (Van Nostrand, New York).

BORN, M., [1933], Optik (Springer, Berlin).

BRILLOUIN, L., [1947], Wave Propagation in Periodic Structures (McGraw-Hill, New York), 1st edition [1938].

BRILLOUIN, L., [1949], Les Tenseurs en Mécanique et en Elasticité (Masson, Paris).

COURANT, R. and D. HILBERT, [1931], Methoden der Mathematischen Physik I (Springer, Berlin).

DEBYE, P., [1925], Physica **5** 377.

DORGELO, H. and J. A. SCHOUTEN, [1946], Proc. Roy. Ac. Amsterdam **48** 124, 282, 393.

EDDINGTON, A. S., [1923], Math. Theory of Relativity (Cambridge) § 74.

EYRING, H., J. WALTER and G. E. KIMBALL, [1944], Quantum Chemistry (Wiley, New York), Chapter XVII.

FOKKER, A. D., [1927], Relativiteits Theorie (Noordhoff, Groningen) in Dutch.

GUILLEMIN, E. A., [1949], The Mathematics of Circuit Analysis (Wiley, New York).

HERTZ, H., [1890], Ann. Phys. Chem. **41** 389.

HILGEVOORD, J., [1960], Dispersion Relations and Causal Description (North-Holland Publ. Co., Amsterdam).

HUBER, A., [1926], Phys. Zeitschr. **27** 619.

LANDAU, L. D. and E. M. LIFSCHITZ, [1957], Electrodynamics (Gostekhizdat Moscow) in Russian; [1960], in English (Addison-Wesley, Reading Mass.).

LEVI CIVITA, T., [1926], The Absolute Differential Calculus (Blackie & Son, London).

LORENTZ, H. A., A. EINSTEIN and H. MINKOWSKI, [1922], Das Relativitätsprinzip, 4th edition (Teubner, Leipzig).

MADELUNG, E., [1953], Mathematische Hilfsmittel des Physikers, 5th edition (Springer, Berlin).

MAXWELL, J. C., [1892], A Treatise on Electricity and Magnetism (Oxford), 1st edition [1873].

MØLLER, C., [1952], The Theory of Relativity (Oxford).

MORSE, P. M. and H. FESHBACH, [1953], Methods of Theoretical Physics, Vol. I (McGraw-Hill, New York).

POLDER, D., [1949], Phil. Mag. **40** 99.

POST, E. J., [1960], Phys. Rev. **118** 1113.

ROSENFELD, L., [1951], Theory of Electrons (North-Holland Publ. Co., Amsterdam).

SCHOUTEN, J. A., [1951], Tensor Analysis for Physicists (Oxford).

SCHOUTEN, J. A., [1954], Ricci Calculus, 2nd edition (Springer, Berlin).

STRATTON, J. A., [1941], Electromagnetic Theory (McGraw-Hill, New York).

SYNGE, J. L., [1956], Relativity, the Special Theory (North-Holland Publ. Co., Amsterdam).

TELLEGEN, B. D. H., [1948], Philips Res. Rep. **3** 81.

TITCHMARSH, E. C., [1948], Introduction to the Theory of Fourier Integrals (Oxford), 1st edition [1937].

TOLMAN, R., [1934], Relativity, Thermodynamics and Cosmology (Oxford).

VAN DANTZIG, D., [1934], Proc. Royal Ac. Amsterdam **37** 526.

VAN VLECK, G. H., [1932], Electric and Magnetic Susceptibilities (Oxford) p. 279.

VEBLEN, O. and J. H. C. WHITEHEAD, [1932], The Foundation of Differential Geometry (Cambridge).

VOIGT, W., [1910], Lehrbuch der Kristallphysik (B. Teubner, Leipzig).

VON LAUE, M., [1920], Phys. Z. 659.

WEYL, H., [1921], Raum, Zeit, Materie, 4th edition (Springer, Berlin); in English (Dover, New York).

WHITTAKER, E. T., [1928], Cambridge, Phil., Tr., **21** 32.

WHITTAKER, E. T., [1953], History of the Theories of Aether and Electricity, Volume II (Nelson, London).

ZOCHER, H. and C. TÖRÖK, [1953], Proc. Nat. Ac. U.S.A. **39** 68.

INDEX

A CATALOG OF SELECTED

DOVER BOOKS

IN SCIENCE AND MATHEMATICS